Contents

Foreword

The goal of these booklets is to provide a problem solving training ground starting from the earliest years of a student's mathematical development.

In our experience, we have found that teaching how to solve problems should focus not only on finding correct answers but also on finding better solution strategies. While the correct answer to a problem can typically be obtained in several different ways, not all these ways are equally useful for learning how to solve problems.

The most basic strategy is *brute force*. For example, if a problem asks for the number of ways Lila and Dina can sit on a bench, it is easy to write down all the possibilities: Dina, Lila and Lila, Dina. We arrive at this solution by performing all the possible actions allowed by the problem, leaving nothing to the imagination. For this last reason, this approach is called brute force.

Obviously, if we had to figure out the number of ways 30 people could stand in a line, then brute force would not be as practical, as it would take a prohibitively long time to apply.

Using brute force to obtain the correct answer for a simpler problem is not necessarily a useful learning experience for solving a similar problem that is more complex. Moreover, solving problems in a quantitative manner, assuming that the student can transfer simple strategies to similar but more complex problems, is not an efficient way of learning problem solving.

From this simple example, we see that the goal of *practicing* problem solving is different from the goal of problem solving. While the goal of problem solving is to obtain a correct answer, the goal of practicing problem solving is to acquire the ability to develop strategies, generate ideas, and combine approaches that are powerful enough to solve the problem at hand as well as future similar problems.

While brute force is not a useless strategy, it is not a key that opens every door. Nevertheless, there are problems where brute force can be a useful tool. For instance, brute force can be used as a first step in solving a complex problem: a smaller scale example can be approached using brute force to help the problem solver understand the mechanics of the problem and generate ideas for solving the larger case.

All too often, we encounter students who can quickly solve simple problems by applying brute force and who become frustrated when the solving methods they have been employing successfully for years become inefficient once problems increase in complexity. Often, neither the student nor the parent has a clear understanding of why the student has stagnated at a certain level. When the only arrows in the quiver are guess-and-check and brute force, the ability to take down larger game is limited.

Our series of books aims to address this tendency to continue on the beaten path - which usually generates so much praise for the gifted student in the early years of schooling - by offering a challenging set of questions meant to build up an understanding of the problem solving process. Solving problems should never be easy! To be useful, to represent actual training, problem solving should be challenging. There should always be a sense of difficulty, otherwise there is no elation upon finding the solution.

Indeed, practicing problem solving is important and useful only as a means of learning how to develop better strategies. We must constantly learn and invent new strategies while questioning the limitations of the strategies we are using. Obtaining the correct answer is only the natural outcome of having applied a strategy that worked for a particular problem in the time available to solve it. Obtaining the wrong answer is not necessarily a bad outcome; it provides insight into the fallacies of the method used or into the errors of execution that may have occured. As long as students manifest an interest in figuring out strategies, the process of problem solving should be rewarding in itself.

Sitting and thinking in a focused manner is difficult to train, particularly since the modern lifestyle is not conducive to adopting open-ended activities. This is why we would like to encourage parents to pull back from a quantitative approach to mathematical education based on repetition, number of completed pages, and the number of correct answers. Instead, open up the time boundaries that are dedicated to math, adopt math as a game played in the family, initiate a math dialogue, and let the student take his or her time to think up clever solutions.

Figuring out strategies is much more of a game than the mechanical repetition of stepwise problem solving recipes that textbooks so profusely provide, in order to "make math easy." Mathematics is not meant to be easy; it is meant to be interesting.

Solving a problem in different ways is a good way of comparing the merits of each method - another reason for not making the correct answer the primary goal of the activity. Which method is more labor intensive, takes more time or is more prone to execution errors? These are questions that must be part of the problem solving process.

In the end, it is not the quantity of problems solved, the level of theory absorbed, or the number of solutions offered in ready-made form by so many courses and camps, but the willingness to ask questions, understand and explore limitations, and derive new information from scratch, that are the cornerstones of a sound training for problem solvers.

These booklets are not a complete guide to the problem solving universe, but they are meant to help parents and educators work in the direction that, aside from being the most efficient, is the more interesting and rewarding one.

The series is designed for mathematically gifted students. Each book addresses an age range as some students will be ready for this content earlier, others later. If a topic seems too difficult, simply try it again in a couple of months.

RATIOS AND PROPORTIONS

Numerically, a ratio is the same number as a fraction. Conceptually, ratios are used to compare quantities that *have the same physical nature* and are *expressed in the same unit*. We cannot compare mass and volume, so there is no such thing as a ratio of mass to volume. But we can certainly define density, which is the result of dividing the mass of an object by its volume. Density is measured in grams per cubic centimeter. By contrast, ratios are *dimensionless*.

A **ratio** compares two quantities expressed in the same unit. The ratio shows what we have to multiply the second quantity (*consequent*) by in order to obtain the first quantity (*antecedent*).

We can say: "the populations of San Francisco and of Ankara are in a ratio of $1 : 5$." But it does not make sense to say: "the population of Paris is in a ratio of $112,000 : 31$ to the height of the Eiffel Tower."

This is an example of comparing two masses:

$$\frac{30 \text{ kg}}{20 \text{ kg}} = \frac{3}{2} = 3 : 2$$

By contrast, this is an example of an average speed:

$$s = \frac{300 \text{ km}}{5 \text{ hr}} = 60 \text{ km/h}$$

When using ratios, it may be necessary to convert units. What is the ratio betwen the time we spent in the car, 2 hours, and the duration of one day? To write this ratio we have to convert 1 day into 24 hours:

$$2 : 24 = 1 : 12$$

Property: A ratio is generally not commutative.

Property: Both terms of a ratio may be multiplied by the same factor to obtain an equivalent ratio:

$$3 : 4 \ = \ 12 : 16$$

$$20 : 30 \ = \ 2 : 3$$

A **proportion** is a pair of equivalent ratios. For example:

$$\frac{3}{5} = \frac{12}{20}$$

Outer terms (or extremes) of a proportion:

$$\frac{\bigcirc}{\cdots} = \frac{\cdots}{\bigcirc}$$

Inner terms (or means) of a proportion:

$$\frac{\cdots}{\bigcirc} = \frac{\bigcirc}{\cdots}$$

Multiplication of means and extremes - the product of the inner terms is equal to the product of the outer terms:

$$\frac{x}{a} = \frac{y}{b} \longrightarrow xb = ya$$

The inner terms may be interchanged:

$$\frac{x}{a} = \frac{y}{b} \longrightarrow \frac{x}{y} = \frac{a}{b}$$

The outer terms may be interchanged:

$$\frac{x}{a} = \frac{y}{b} \longrightarrow \frac{b}{a} = \frac{y}{x}$$

Adding the denominators of a proportion to the numerators results in a new set of 4 proportional numbers. If

$$\frac{a}{b} = \frac{x}{y}$$

then it is also true that

$$\frac{a+b}{b} = \frac{x+y}{y}$$

because

$$\frac{a}{b} + 1 = \frac{x}{y} + 1$$

In this way, from a set of 4 proportional numbers:

$$a, \; b, \; x, \; y$$

we derived another set of 4 proportional numbers:

$$a+b, \; b, \; x+y, \; y$$

Subtracting the denominators of a proportion from the numerators results in a new set of 4 proportional numbers. If

$$\frac{a}{b} = \frac{x}{y}$$

then

$$\frac{a-b}{b} = \frac{x-y}{y}$$

because

$$\frac{a}{b} - 1 = \frac{x}{y} - 1$$

In this way, from a set of 4 proportional numbers:

$$a, \ b, \ x, \ y$$

we derived another set of 4 proportional numbers:

$$a - b, \ b, \ x - y, \ y$$

Multiple Ratios are equalities between several equivalent ratios.

In a multiple ratio, the sum of the numerators and the sum of the denominators form a ratio equivalent to the other ratios:

$$\frac{3}{5} = \frac{9}{15} = \frac{12}{20} = \frac{3+9+12}{5+15+20}$$

More properties of multiple ratios

$$\frac{a}{b} = \frac{m}{n} = \frac{p}{q} = \frac{a+m+p}{b+n+q} = r$$

leads to:

$$\frac{aw}{bw} = \frac{mv}{nv} = \frac{pz}{qz} = \frac{aw+mv+pz}{bw+nv+qz} = r$$

CHAPTER 2. RATIOS AND PROPORTIONS

Dividing a quantity in a given ratio:

The quantity Q must be divided into three parts that have to be in the ratio $n : m : p$.

Denote the parts by a, b, c. Then, $a + b + c = Q$, and

$$\frac{a}{n} = \frac{b}{m} = \frac{c}{q}$$

Denote the value of this ratio by R:

$$\frac{a}{n} = \frac{b}{m} = \frac{c}{q} = R$$

Then

$$a = nR \qquad b = mR \qquad c = qR$$

$$a + b + c = Q = nR + mR + qR = R(n + m + q)$$

$$R = \frac{Q}{n + m + q}$$

Example

Divide 325 into three parts that are in a ratio of $3 : 4 : 6$.

The number of parts is $3 + 4 + 6 = 13$. The size of each part is $325 \div 13 = 25$. The parts are:

$$
\begin{aligned}
3 \times 25 &= 75 \\
4 \times 25 &= 100 \\
6 \times 25 &= 150
\end{aligned}
$$

PRACTICE ONE

Do not use a calculator for any of the problems!

Exercise 1

A proportion is formed with the numbers 3, 12, 5, and x. x cannot be:

(A) 1.25

(B) 4

(C) 7.2

(D) 20

Exercise 2

Three numbers, a, b, and c, are in the same ratios as $4, 5$, and $\dfrac{7}{3}$. If their sum is 204, what is the value of $a - c$?

(A) 18

(B) 26

(C) 30

(D) 31

Exercise 3

The quantities of vinegar and oil in the common vinaigrette sauce are in a ratio of $1.5 : 2$. Dina wants to prepare 140 mL of vinaigrette. How much vinegar must she use?

Exercise 4

Ali and Baba have amassed a considerable fortune in gold and silver coins by looting caves in the desert. The ratio of silver coins to gold coins is 5 : 3. The mass of one silver coin and the mass of one gold coin are in a ratio of 3 : 2. Ali and Baba made some calculations and found out that they could buy just as much with their silver coins as with their gold coins. What is the ratio between the value of a gram of silver and the value of a gram of gold?

Exercise 5

Ali, Baba, and their friend Ulug played a game. They each started with a number of identical silver coins. After a few rounds, Ali won 5 coins and Ulug won 8 coins. At that time, the numbers of coins each of them had were in the ratio 3 : 5 : 4, respectively. If they had 360 coins in total, how many coins did each of them start with?

Exercise 6

Alfonso, the grocer, started a local veggie delivery program. One day, he filled the delivery basket with leeks, peppers, and beets in a ratio of 3 : 2 : 5. A few days later, due to fluctuations in supply, the ratio became 1 : 2 : 3. If Alfonso always packs the same total number of vegetables in each basket, what fraction of the old number of leeks is the new number of leeks?

Exercise 7

The sum of four numbers is 301. The numbers are in the ratios 15 : 11 : 8 : 9. What are the numbers?

Exercise 8

3300 tons of fuel must be delivered to three neighborhoods. There are 250 inhabitants in one neighborhood, 500 in another, and 350 in the last. If the fuel is divided equally among inhabitants, what quantity of fuel will be delivered to each neighborhood?

Exercise 9

Three positive integers are in the ratios $2 : 4 : 7$. The difference between the largest and the smallest numbers is 55. What are the numbers?

Exercise 10

Write proportions using the numbers: 7, 11, 56, and 88.

Exercise 11

Which of the following sets of 4 numbers can form proportions? Assume $a \neq b$ and $m \neq n$. Check all that apply.

(A) $\{52, \ 8, \ 13, \ 32\}$

(B) $\{52, \ 4, \ 13, \ 39\}$

(C) $\{11, \ 88, \ 28, \ 7\}$

(D) $\{2a, \ 5a, \ 10b, \ b\}$

(E) $\{m+1, \ n+1, \ m-1, \ n-1\}$

Exercise 12

On a map, the distance between two points is represented by an 8 mm long segment. The scale of the map is $\frac{1}{2000000}$. What is the actual distance, in kilometers, between the two points?

Exercise 13

A wheel spins 100 times in 2 minutes and 30 seconds. How many times does it spin in 15 minutes?

Exercise 14

If the real distance between points A and B is 300 miles, what is the distance, in inches, that separates them on a map drawn at a scale of $1 : 1500000$? Calculator allowed.

Exercise 15

If x and y are terms in the proportion:

$$\frac{3}{x} = \frac{y}{5}$$

find the value of the expression:

$$E = \frac{xy + 7}{2xy - 8}$$

Exercise 16

Find the value of the ratio $p : q$, knowing that:

$$\frac{8p}{4q - 5p} = 11$$

Exercise 17

If $p : q$ is the same as $3 : 4$, compute the value of the expression:

$$F = \frac{5q + p}{5q - 3p}$$

Exercise 18

Divide the number 625 in three parts that are in the same ratios as the numbers 2, 3, and $3.\overline{3}$.

Exercise 19

If the time left until the end of the day is $\frac{3}{7}$ of the time that has already passed since the day began, what time is it now?

Exercise 20

A path is 175 feet shorter than another path. One third of the length of one path is equal to three fourths of the length of the other path. Find the lengths of the two paths.

PERCENTS

A *percent* is a fraction with denominator 100. Few fractions can be rewritten exactly to have a denominator of 100. However, percentages are used in practice because they provide a unified way of comparing quantities.

The ratio 1 : 5 can be rewritten as 20 : 100 and *denoted* further as 20%. Note that:

$$20\% = \frac{20}{100}$$

Therefore,

$$.1234 = 12.34\%$$

Experiment

Write each decimal or fraction as a percentage:

$$0.235 \ =$$

$$1.237 \ =$$

$$\frac{9}{5} \ =$$

$$0.015 \ =$$

$$\frac{11}{25} \ =$$

$$\frac{91}{1000} \ =$$

Answers:

23.5% 123.7%, 180%, 1.5%, 44%, 9.1%

Examples of exact percentages and their representations as irreducible fractions and as decimals:

Fraction	Percentage	Decimal
$\frac{1}{10} = \frac{10}{100}$	10%	0.1
$\frac{1}{5} = \frac{20}{100}$	20%	0.2
$\frac{1}{4} = \frac{25}{100}$	25%	0.25
$\frac{1}{2} = \frac{50}{100}$	50%	0.5
$\frac{3}{4} = \frac{75}{100}$	75%	0.75

Finding $p\%$ of a quantity Q

$$x : Q \;=\; p : 100$$

$$\frac{x}{Q} \;=\; \frac{p}{100}$$

$$x \cdot 100 \;=\; p \cdot Q$$

$$x \;=\; \frac{p}{100} \cdot Q$$

Finding a quantity Q if $p\%$ of it is known

$$x : Q \;=\; p : 100$$

$$\frac{x}{Q} \;=\; \frac{p}{100}$$

$$Q \;=\; \frac{x \cdot 100}{p}$$

Finding the percentage of a quantity Q that another quantity x represents

$$x : Q \;=\; p : 100$$

$$\frac{x}{Q} \;=\; \frac{p}{100}$$

$$p \;=\; \frac{x \cdot 100}{Q}$$

$$\;=\; \frac{x}{Q} \cdot 100$$

Percent increase/decrease

$$\frac{Q_{\text{final}} - Q_{\text{initial}}}{Q_{\text{initial}}} \cdot 100\%$$

Final amount after percent increase/decrease

After an increase of $p\%$, a quantity Q becomes:

$$Q_{\text{new}} = Q_{\text{old}} + Q_{\text{old}} \times \frac{p}{100}$$

$$= Q_{\text{old}}\left(1 + \frac{p}{100}\right)$$

After a decrease of $p\%$, a quantity Q becomes:

$$Q_{\text{new}} = Q_{\text{old}} - Q_{\text{old}} \times \frac{p}{100}$$

$$= Q_{\text{old}}\left(1 - \frac{p}{100}\right)$$

Examples:

1. The amount N increases by 15%. What is the final amount?

2. The amount N decreases by 15%. What is the final amount?

3. The amount N increases by 30%. What is the final amount?

4. The amount N decreases by 30%. What is the final amount?

Answers:

1. $N_{\text{new}} = N_{\text{old}} \times 1.15$

2. $N_{\text{new}} = N_{\text{old}} \times 0.85$

3. $N_{\text{new}} = N_{\text{old}} \times 1.3$

4. $N_{\text{new}} = N_{\text{old}} \times 0.7$

Repeated application of percent increases/decreases

A quantity to which repeated percent increases and/or decreases are applied will be affected independently by each change. This is a consequence of the commutative property of multiplication.

Example: A quantity of 100 is decreased by 15%. The result is increased by 20% and then decreased by 5%. What is the final quantity?

$$Q_{\text{new}} = Q_{\text{old}} \times 0.85 \times 1.2 \times 0.95$$

$$= Q_{\text{old}} \times 0.969$$

$$Q_{\text{new}} = 100 \times 0.969$$

$$Q_{\text{new}} = 96.9$$

PRACTICE TWO

Exercise 1

A quantity is increased by 10%, then by 15%, and is finally decreased by 20%. Overall, did the quantity increase or decrease? What is the percent increase/decrease?

Exercise 2

Two quantities are in a ratio of 5 : 4. If the first quantity increased by 20%, to which of the following did its ratio to the other quantity change?

(A) $3 : 2$

(B) $4 : 5$

(C) $6 : 5$

(D) $1.2 : 5$

Exercise 3

Due to changes in the design of the Mouse Maze, the path to the Cheese has been shortened by 10%. At the same time, a diligent workout routine has increased the Mouse's speed by 25%. By how much percent has the average time of access to the Cheese changed? Is it an increase or a decrease?

Exercise 4

In modern English, 29% of the vocabulary is assumed to be of French origin and 26% of Germanic origin. If the vocabulary grows by 0.5% each decade by adopting words of a different origin, will the difference between the percentage of French-origin words and the percentage of Germanic-origin words be more than or less than 3% after 2 decades?

Exercise 5

The width of a rectangle increased by 60% and its length increased by 20%. By what percentage did its area increase?

Exercise 6

The width of a rectangle increased by 30% and its length decreased by 30%. Did its area increase or decrease? By what percent?

Exercise 7

A circle's radius increased by 5%. By what percentage did its area increase?

Exercise 8

What percent of N is 10% of 120% of 35% of N?

(A) 3.26%

(B) 4.2%

(C) 40%

(D) 42%

Exercise 9

If the 30° angle of a triangle increases by 50%, by what percentage does the sum of the other two angles decrease?

Exercise 10

The sum of three numbers is 87. If we increase the first number by 150%, decrease the second one by 25%, and decrease the third one by 5, the results are equal. Find the numbers.

Exercise 11

Alfonso sorted 420 lbs of potatoes in two containers so that 30% of the contents of one container have the same weight as 40% of the contents of the other container. The difference between the amounts of potatoes in the two containers is:

(A) 42 lbs

(B) 50 lbs

(C) 60 lbs

(D) 70 lbs

Exercise 12

What percentage of the positive integers between 34 and 147 are multiples of 8?

Exercise 13

The side of a square increased by 50%. By what percentage did its area increase?

Exercise 14

Due to drought conditions, only a third of Dina's tomato plants survived last week's heat. What was the percent decrease in tomato plants?

Exercise 15

A store uses a profit/loss simulator to determine the effect of multiple rebates in prices. The store manager inputs the products he wants to put on sale and applies a rebate of 15%. The simulator shows a profit of 25% for that set of products. The manager applies another rebate to the same set of products and the profit goes down to 0%. Find the value of the second rebate.

MIXTURES

Problems with *mixtures* typically involve the application of a law of conservation: mass, volume, number of particles.

Mixtures are characterized by an *initial* and a *final* state. Between these two states, the concentration or the relative number of particles change, but the mass or the total number of particles remains the same (is *invariant*).

To solve a problem with mixtures, it is useful to set up an equation that reflects the conservation of some physical quantity between the initial and final states.

Example: Lynda, the terrier, and Arbax, the Dalmatian, have decided to merge their treat stashes. Lynda's stash consisted of 40% Veg-a-Dog treats and 60% Vit-a-Dog treats. Arbax's stash consisted of 25% Veg-a-Dog treats and a variety of mixed treats. If Arbax's stash contained twice as many treats as Lynda's stash, what percentage of the merged collection is made up of Veg-a-Dog treats?

Denote Lynda's treats by L and Arbax's treats by $2 \times L$. The merged collection contains $3 \times L$ treats.

Set up an equation between the total number of Veg-a-Dog treats before the merger:

$$L \times \frac{40}{100} + 2 \times L \times \frac{25}{100}$$

and after the merger (x is the percentage we are looking for):

$$3 \times L \times \frac{x}{100}$$

These amounts must be equal:

$$L \times \frac{40}{100} + 2 \times L \times \frac{25}{100} = 3 \times L \times \frac{x}{100}$$

Multiply the entire equation by 100 and divide it by L:

$$
\begin{aligned}
40 + 50 &= 3 \times x \\
90 &= 3 \times x \\
x &= 30
\end{aligned}
$$

The percentage of Veg-a-Dogs in the merged stash is 30%.

PRACTICE THREE

Do not use a calculator for any of the problems!

Exercise 1

When we add 100 grams of water to a water and salt solution that has a salt concentration of 20%, the salt concentration drops to 16%. What is the final mass of the solution?

Exercise 2

Stephan had a box of tennis balls. 80% of the balls were yellow and the rest were green. What percentage of the green balls must be replaced by yellow balls so that the percentage of yellow balls in the box will become 85%?

Exercise 3

Dina and Lila made a fruit punch that is 50% juice and 50% water. They drank half of the punch and refilled the jug with water. Then, they drank half of the punch again and refilled the jug with juice. What was the final concentration of the juice?

Exercise 4

Ali and Baba have a fortune in gold and silver coins. If they spent 600 gold coins and 200 silver coins, the remaining gold coins would represent 55% of the total number of coins. If they spent 200 gold coins and 400 silver coins, the remaining gold coins would represent 65% of the total number of coins. How many coins do Ali and Baba have?

Exercise 5

A mixture of two copper ores, one that is 35% copper and another that is 42% copper, has a copper concentration of 40%. If we add 350 kilograms of the less concentrated ore to the mixture, the copper concentration in the mixture drops to 38%. What is the total mass, in kilograms, of the original mixture?

Exercise 6

A two liter solution consists of 95% water and 5% vinegar. How much water should we add to it so that the vinegar concentration falls to 2%?

Exercise 7

Dina dissolved some sugar in water so that the solution was 12% sugar by mass. Lila did the same but her solution was 18% sugar by mass. They poured both solutions into a pitcher and added another 150 g of water. The pitcher now contained 1200 g of solution with a sugar concentration of 15%. How many grams of solution did each of them have initially?

Exercise 8

Amira has a bag of blocks. 60% of them are green and the rest are blue. If Amira paints 30% of the green blocks blue and 40% of the blue blocks green, what percentage of the blocks will be green?

Exercise 9

How many mL of pure alcohol do we have to mix with a 28% concentrated solution of alcohol in water in order to obtain 360 mL of solution that is 35% concentrated?

Exercise 10

We have 200 mL of pure alcohol and 1200 mL of 15% concentrated alcohol in water. What is the largest amount, in mL, of 40% concentrated solution of alcohol in water we can make?

RATES

A **rate** shows how a quantity changes over time. In problems at this level, we assume that all rates are constant. This means that objects never "speed up" or "slow down." Instead, objects always travel at the same speed. Unless otherwise specified, objects are considered to be *pointlike*. This means they have no size of their own.

Since problems with rates handle *physical quantities and not just numbers*, we have to make sure we compare them using the same unit! In problems with rates, it is sometimes necessary to *change the units* to make them all the same. Before attempting the problems in this section, students should review *metric system conversions*.

Speed is defined as the rate at which distance changes over time. For problems about motion (i.e. *kinematics*), distance, average speed, and time are related as follows:

$$\text{Distance} = \text{Speed} \times \text{Time}$$

The distance and the speed vary *directly*: the higher the speed, the greater the distance.

The distance and the time vary *directly*: the longer the time, the greater the distance.

The speed and the time vary *inversely*: the longer the time, the lower the speed.

Meeting Problems involve moving objects that meet. The objects are, in general, considered to be *pointlike*. This means that they have no dimensions of their own.

If two moving objects meet, then they are *at the same place at the same time*. The solution hinges on the correct application of this simple fact.

Example 1: A horse and a bicycle start at either end of a trail and arrange to meet midway. The horse travels at 2 m/s and the bicycle travels at 2.5 m/s. The bicycle reaches the meeting point 12 minutes before the horse does. How long is the trail, in km?

Denote the time, in seconds, from when they started until they met by T. Convert the 12 minutes to seconds and subtract them from T to find out how long it took the bicycle to reach the meeting point.

We know the distances they covered were both equal to half the length of the trail. This equation sets the two distances to be equal:

$$2 \times T = 2.5 \times (T - 12 \times 60)$$

$$2 \times T = 2.5 \times T - 2.5 \times 12 \times 60$$

$$0.5 \times T = 1800 \text{ s}$$

$$T = 3600 \text{ s} = 60 \text{ min} = 1 \text{ hr}$$

Since $2 \, \dfrac{\text{m}}{\text{s}} \times T$ is equal to half the length of the trail, the entire length, in km, is:

$$L = 2 \times 2 \times 60 \times 60$$

$$L = 14400 \text{ m}$$

$$L = 14.4 \text{ km}$$

The **average speed** is:

$$\text{Speed} = \frac{\text{Total Distance}}{\text{Total Time}}$$

Example: A vehicle travels one third of the way at an average speed of 50 mph and the remaining two thirds at an average speed of 60 mph. What was the overall average speed of the vehicle?

Denote the distance by $3d$. One third of the distance is d and the remaining is $2d$. The travel time for the first part of the journey is:

$$t_1 = \frac{d}{50}$$

and the travel time for the second part of the journey is:

$$t_2 = \frac{2d}{60}$$

The average speed is computed as follows:

$$
\begin{aligned}
\text{Speed} &= \frac{3d}{\frac{d}{50} + \frac{2d}{60}} \\[2ex]
&= \frac{3d}{\frac{6d+10d}{300}} \\[2ex]
&= \frac{3d}{1} \times \frac{300}{16d} \\[2ex]
&= \frac{900}{16} = 56.25 \text{ mph}
\end{aligned}
$$

Note: This is different from the average of the two average speeds.

$$\frac{50 + 60}{2} = 55 \text{ mph}$$

Using Reduction to Unity to Solve Rate Problems

Example 1

40 walnut trees produce 3600 lbs of nuts over a 6 year span. How many walnut trees produce 2040 lbs of nuts over an 8 year span?

years	walnuts	production	description
6	40	3600	40 trees produce 3600 lbs in 6 years
1	40	600	40 trees produce 600 lbs in 1 year
1	1	15	1 tree produces 15 lbs in 1 year
8	1	120	1 tree produces 120 lbs in 8 years
8	17	2040	17 trees produce 2040 lbs in 8 years

Experiment

60 deer graze 540 lbs of grass over a 3 hour interval. How many deer will graze 300 lbs of grass over a 2 hour interval? Fill in the missing values:

deer	60	60	1	1	z
hours	3	1	1	y	2
lbs grass	540	x	3	6	300

Answer: $x = 180,$ $y = 2,$ $z = 50.$

PRACTICE FOUR

Do not use a calculator for any of the problems!

Exercise 1

A team of 3 masons can finish a job in 5 days if they work 8 hours per day. How many days will it take 4 masons, working only 6 hours per day, to complete the same job? (Assume all the masons work at the same rate.)

Exercise 2

To empty a pool, we can use 3 taps. If the first tap is open for 2 hours, the second tap is open for 3 hours, and the third tap is open for 6 hours, 22000 liters of water are going to flow out. If we run the first tap for 3 hours, the second tap for 2 hours, and the third tap for 6 hours, 21000 liters of water will flow out. If the first and the second taps are run for 2 hours and the third tap for 3 hours, 14500 liters of water flow out. How many liters flow through each of the taps in one hour?

Exercise 3

A crew of workers is using two types of hoses to fill a container. Water flows through one type of hose at a rate of 250 liters/hour and through the other type at a rate of 270 liters/hour. In an hour, 1060 liters of water flow into the container. How many hoses of each type are there?

Exercise 4

A railroad is due for maintenance. If 12 workers can repair half the length in 28 days, how many days will it take to repair the other half if 4 workers quit?

Exercise 5

A 10 person team completes a task in 18 days. Another 18 people team completes the same task in 12 days. How long will it take a team formed of 5 people from the first group and 12 people from the second group to finish the job? (Assume that the workers in each group work at the same rate, and that a worker in group 1 works at a different rate from a worker in group 2.)

Exercise 6

Two faucets can fill a tub if they run simultaneously for 12 hours. After 8 hours of simultaneous operation, one of the faucets is shut. It took the other faucet 10 hours to fill the tub completely. What time would it take each faucet to fill the empty tub alone?

Exercise 7

Dina and Amira's families rented a house to go on vacation. They paid 1400 dollars in total. Amira's family has 3 members and stayed for 5 days. Dina's family has 5 members and stayed for 4 days. How should the two families share the expenses fairly?

Exercise 8

Dina and Lila have some old watches that do not work properly anymore. Dina's watch advances by 4 minutes every hour and Lila's watch falls behind by 6 minutes every hour. If they both set their watches to show the same time in the morning, by evening Dina's watch shows 7:36 PM while Lila's watch shows 6:06 PM. At what time did they set their watches to show the same time?

Exercise 9

Ali and his donkey start walking from Ali's house towards the Date Palm Oasis. Ali walks three times faster than the donkey. He reaches the oasis, fills a bag with dates and starts back towards home. After leaving the oasis, Ali meets his donkey one third of the way home and decides to ride him all the way home. If it took Ali 15 minutes to fill the bag, how long will it take them to get home from the meeting point?

Exercise 10

Dina and Lila went hiking in the park with their mother. They were driving home when their mother had to stop for a train. Between the time the locomotive passed them and the time the last car passed them, they waited at the barrier for 1.5 minutes. Farther on, the train had to pass through a 1.5 mile long tunnel which it did in 6 minutes. If the speed of the train was constant, what was the length of the train?

Exercise 11

The little train at the Zoo runs in a loop that is 40 times its length. Assuming the train never stops and runs at a constant speed, what percentage of the time is the train directly in front of the cashier's booth?

Exercise 12

Stephan is a member of a sailing team. In a race, they had to sail from buoy A to B to C to D, and then sail back on the same route. They sailed from A to B at a speed of 9 knots, from B to C at a speed of 4 knots, and from C to D at a speed of 3 knots. On the way back, the speeds were exactly reversed. The way from A to D took twice as long as the way from D to A. Which of the following could be the ratio AB:BC:CD?

(A) $1 : 4 : 14$

(B) $9 : 4 : 3$

(C) $3 : 4 : 9$

(D) $1 : 7 : 13$

Exercise 13

A vehicle's speed changes as follows in a 2 hour time interval. What is the distance covered by the vehicle during this time?

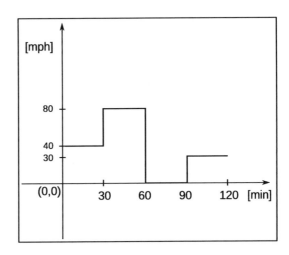

Exercise 14

A vehicle's speed changes as follows in a 2 hour time interval. What is the distance covered by the vehicle during this time?

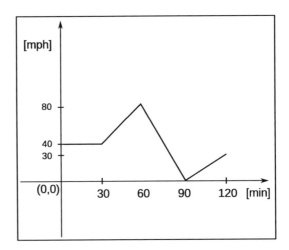

Exercise 15

Stephan had to travel to a tennis tournament. He traveled by plane for 4 hours at an average speed of 600 mph and then by car for 2 hours at an average speed of 75 mph. What was his average speed overall?

Exercise 16

A mouse positioned at corner A of a square runs along the perimeter of the square on the shortest path towards point P, the midpoint of BC. A snail positioned in the center of the square at point O crawls towards P on the shortest possible path. If they leave at the same time and arrive at P simultaneously, what is the ratio between the speed of the snail and the speed of the mouse?

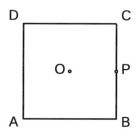

(A) $3:1$

(B) $2:1$

(C) $1:2$

(D) $1:3$

Exercise 17

Dina and Lila wanted to go to a Hallowe'en party but they only had one broom. So they decided that Dina could use the broom while Lila walked and that Dina would leave the broom somewhere along the way for Lila to find and use to get to their destination. Both of them walked at 2 miles per hour and rode the broom at 5 miles per hour. If Lila arrived 36 minutes before Dina and the broom remained unused for 18 minutes, how far from the party is their house?

MISCELLANEOUS PRACTICE

Do not use a calculator for any of the problems!

Exercise 1

If the length of a rectangle is increased by 0.5 cm, its area increases by 2.5 cm². If the width of the same rectangle is increased by 0.25 cm, its area increases by 1.25 cm². Prove that this rectangle is a square.

Exercise 2

A metal container is filled with liquid to $x\%$ of its total volume. The container is heated. As a result of thermal expansion, the volume of the container increases by 8% and the volume of the liquid increases by 20%. What must x be if the container is now completely full?

Exercise 3

96% of the volume of a cube is filled with water. If all the sides of the cube double in length, what percentage of the new volume will be filled with water? (Assume the amount of water does not change.)

Exercise 4

A flock of ducklings were crossing a river. 10 of them swam to the other side of the river and half of the remaining ducklings followed them. At that point, there were one third as many ducklings that had not crossed as there were ducklings that crossed. How many ducklings were there in total?

Exercise 5

In two containers, one glass and one plastic, there are 26 liters of grape juice. If we transfer 3 liters from the glass container to the plastic container, there are 8 liters less in the glass container than in the plastic container. How many liters of juice were there in the glass container to start with?

Exercise 6

Stephan was driving to a tennis tournament. After he drove $\frac{2}{7}$ of the total route, he found that were 150 miles left to cover before getting halfway to destination. How long was Stephan's route?

Exercise 7

The numbers a, b, and c are in the ratios $4 : 7 : 9$. Compute the ratio:

$$T = \frac{a + 3b}{b + 2c}$$

Exercise 8

The triangle AMC is obtained by increasing the height AM of triangle ABC by 15%. The triangle ANB is obtained by decreasing the base BC of triangle ABC by 8%. What is the ratio of the areas of triangles AMC and ANB?

Exercise 9

The average of four numbers is x. The numbers are in a ratio of $2 : 3 : 5 : 8$. What is the range of the numbers as a function of x?

(A) $\dfrac{x}{2}$

(B) $\dfrac{3x}{4}$

(C) $\dfrac{4x}{3}$

(D) $\dfrac{2x}{3}$

41

Exercise 10

After two successive price decreases, by 10% and 20% respectively, the price of a car is now 36,000 dollars. What was the price before the rebates?

Exercise 11

A number a is 20% of another number b. A number c is 20% of a and a number d is 60% of c. Which of the following is equal to the sum $a + b + c + d$? Check all that apply.

(A) $6.32 \cdot a$

(B) $1.264 \cdot b$

(C) $31.6 \cdot c$

(D) $52.\overline{6} \cdot d$

Exercise 12

15% of the number of girls in Amira's school is equal to 18% of the number of boys. What is the smallest number of students the school may have?

Exercise 13

The number of girls in Dina's class is equal to 90% of the number of boys. What percent of the number of girls is the number of boys?

Exercise 14

Robot BeeToo tiles a floor using square tiles. Robot DeeFoo tiles an identical floor with tiles that have a side equal to 25% of the side of the tiles used by BeeToo. DeeFoo places tiles at a frequency that is 700% higher than that of BeeToo. Which of the following is true?

(A) In the same time, BeeToo can tile 2 times as much surface as DeeFoo.

(B) In the same time, DeeFoo can tile 2 times as much surface as BeeToo.

(C) It takes DeeFoo 4 times longer than BeeToo to tile the entire floor.

(D) It takes BeeToo 4 times longer than DeeFoo to tile the entire floor.

Exercise 15

Alfonso, the grocer, sold 900 lbs of potatoes in June and 744 lbs of potatoes in July. What was the percent decrease in average daily sales between the two months?

Exercise 16

A delivery truck left the avocado farm to bring a shipment of avocadoes to Alfonso's grocery. The truck was supposed to reach the grocery at 9:00 AM, traveling at an average speed of 50 mph. However, the truck arrived at 10:30 AM because it only managed an average speed of 40 mph. What was the distance between the avocado farm and Alfonso's store?

Exercise 17

Lila and Dina rode a tandem bike. For the first half of the time, they rode at an average speed of 3 mph. For the second half of the time, they rode at an average speed of 4.5 mph. What was their overall average speed?

Exercise 18

Lila and Dina rode a tandem bike. For the first half of the way, they rode at an average speed of 3 mph. For the second half of the way, they rode at an average speed of 4.5 mph. What was their overall average speed?

Exercise 19

A 4×6 rectangle, a 5×2 rectangle, and a 2×3 rectangle are placed so as to form a figure with the smallest possible perimeter, but without overlapping each other. By forming the figure, which of the following is closest to the percent decrease of the total perimeter?

(A) 18%

(B) 24%

(C) 28%

(D) 36%

Exercise 20

A truck and a car leave point P at 8:00 AM and travel towards point Q at 50 mph and 60 mph respectively. A motorcycle leaves point Q at 8:30 AM and travels towards point P at 60 mph on the same road. The motorcycle meets the car and, 12 minutes later, meets the truck. What is the distance between P and Q?

Exercise 21

The sum of three numbers is 111. If we increase the first number by 150%, decrease the second one by 25%, and decrease the third one by 5, the results are equal. Find the numbers.

Exercise 22

The base of a triangle increased by 15% and the corresponding height decreased by 10%. Which of the following is closest to the percent increase or decrease of its area?

(A) 3.5% increase

(B) 5% increase

(C) the area remains the same

(D) 0.5% decrease

Exercise 23

Lila pours 20 mL of milk in a bottle with 200 mL of coffee. Dina pours 20 mL of coffee in a bottle with 200 mL of milk. By how much percent is Lila's mixture more concentrated in coffee than Dina's mixture?

Exercise 24

Ali and Baba were counting their gold coins. Ali said to Baba: "Right now, you have twice as many gold coins as I have. If you give me 205 gold coins and if I give you 305 gold coins, then you will have four times as many gold coins as I have." How many gold coins did Ali have when he made this statement?

SOLUTIONS TO PRACTICE ONE

Do not use a calculator for any of the problems!

Exercise 1

A proportion is formed with the numbers 3, 12, 5, and x. x cannot be:

(A) 1.25

(B) 4

(C) 7.2

(D) 20

Solution 1

The answer is (B). There are 3 possible ways to create equal products:

$$5 \times x = 3 \times 12$$

$$3 \times x = 5 \times 12$$

$$12 \times x = 3 \times 5$$

and the possible values of x are:

$$x = \frac{3 \times 12}{5} = 7.2$$

$$x = \frac{5 \times 12}{3} = 20$$

$$x = \frac{3 \times 5}{12} = 1.25$$

Exercise 2

Three numbers, a, b, and c, are in the same ratios as $4, 5$, and $\dfrac{7}{3}$. If their sum is 204, what is the value of $a - c$?

(A) 18

(B) 26

(C) 30

(D) 31

Solution 2

Use the properties of multiple ratios:

$$\frac{a}{4} = \frac{b}{5} = \frac{3c}{7} = \frac{a + b + 3c}{16}$$

Use the last pair of ratios and the sum of the numbers to calculate c:

$$\frac{3c}{7} = \frac{a + b + c + 2c}{16}$$

$$\frac{3c}{7} = \frac{204 + 2c}{16}$$

$$16 \times 3c = 7 \times (204 + 2c)$$

$$16 \times 3c = 7 \times 204 + 14 \times c$$

$$(48 - 14)c = 7 \times 204$$

$$34c = 7 \times 17 \times 12$$

$$17 \times 2 \times c = 7 \times 17 \times 2 \times 6$$

$$c = 42$$

We can calculate a from the first and last ratios:

$$\frac{a}{4} = \frac{3c}{7}$$

$$7a = 4 \times 3 \times c$$

$$7a = 4 \times 3 \times 7 \times 6$$

$$a = 12 \times 6$$

$$a = 72$$

Then $a - c = 72 - 42 = 30$. The answer is (C).

Exercise 3

The quantities of vinegar and oil in the common vinaigrette sauce are in a ratio of 1.5 : 2. Dina wants to prepare 140 mL of vinaigrette. How much vinegar must she use?

Solution 3

Strategy 1: Write a proportion with the ratios of vinegar to oil. If the total amount of vinaigrette is 140 mL, then the amounts of vinegar and oil can be parametrized as x and $140 - x$, respectively:

$$\frac{1.5}{2} = \frac{x}{140 - x}$$

Cross-multiply and solve for x:

$$\frac{3}{4} = \frac{x}{140 - x}$$

$$3 \times (140 - x) = 4 \times x$$

$$420 - 3x = 4x$$

$$420 = 7x$$

$$x = 60$$

Dina must use 60 mL of vinegar.

Strategy 2: Write the multiple ratio of vinegar, oil, and sauce:

$$1.5 : 2 : 3.5$$

and multiply it by a convenient factor (40) to obtain the total amount of vinaigrette sauce required:

$$60 : 80 : 140$$

Dina must use 60 mL of vinegar and 80 mL of oil.

Exercise 4

Ali and Baba have amassed a considerable fortune in gold and silver coins by looting caves in the desert. The ratio of silver coins to gold coins is 5 : 3. The mass of one silver coin and the mass of one gold coin are in a ratio of 3 : 2. Ali and Baba made some calculations and found out that they could buy just as much with their silver coins as with their gold coins. What is the ratio between the value of a gram of silver and the value of a gram of gold?

Solution 4

The value of the gold is given by the product:

number of gold coins $\times$ mass of a gold coin $\times$ value of one gram of gold

and similarly for the silver.

Denote the value of a gram of sillver with S and that of a gram of gold with G. Then we have:

$$\frac{5 \times 3 \times S}{3 \times 2 \times G} = 1$$

$$\frac{5 \times \cancel{3} \times S}{\cancel{3} \times 2 \times G} = 1$$

$$\frac{5 \times S}{2 \times G} = 1$$

$$\frac{S}{G} = \frac{2}{5}$$

The answer is 2 : 5.

Exercise 5

Ali, Baba, and their friend Ulug played a game. They each started with a number of identical silver coins. After a few rounds, Ali won 5 coins and Ulug won 8 coins. At that time, the numbers of coins each of them had were in the ratio 3 : 5 : 4, respectively. If they had 360 coins in total, how many coins did each of them start with?

Solution 5

If Ali won 5 and Ulug won 8, Baba must have lost 13 coins. Denote the numbers of coins they each started with by A, B, and U, respectively. The multiple ratio is:

$$\frac{A+5}{3} = \frac{B-13}{5} = \frac{U+8}{4}$$

Use the property of adding all numerators and all denominators to obtain an equivalent ratio:

$$\frac{A+5}{3} = \frac{B-13}{5} = \frac{U+8}{4} = \frac{A+B+U}{12}$$

Replace $A + B + U$ with 360:

$$\frac{A+5}{3} = \frac{B-13}{5} = \frac{U+8}{4} = \frac{360}{12} = 30$$

And now solve for each in turn:

$$A + 5 = 90$$
$$A = 85$$

The answer is: $A = 85$, $B = 163$, and $U = 112$.

Exercise 6

Alfonso, the grocer, started a local veggie delivery program. One day, he filled the delivery basket with leeks, peppers, and beets in a ratio of $3 : 2 : 5$. A few days later, due to fluctuations in supply, the ratio became $1 : 2 : 3$. If Alfonso always packs the same total number of vegetables in each basket, what fraction of the old number of leeks is the new number of leeks?

Solution 6

For the first day, denote the number of leeks by L, of beets by B, and of peppers by P. For the last day, denote the number of leeks by l, of beets by b, and of peppers by p. Write the multiple ratios and apply the properties of ratios:

$$\frac{L}{3} = \frac{P}{2} = \frac{B}{5} = \frac{L+P+B}{10}$$

$$\frac{l}{1} = \frac{p}{2} = \frac{b}{3} = \frac{l+p+b}{6}$$

Denote the number of vegetables in the basket by s. It is the same on both days:

$$s = L + B + P$$

$$s = l + b + p$$

Compute the number of vegetables s in two different ways from the previous equalities:

$$s = 10 \times \frac{L}{3} = 6 \times \frac{l}{1}$$

$$10 \times L = 18 \times l$$

$$5 \times L = 6 \times l$$

$$l = \frac{5}{6} \times L$$

Exercise 7

The sum of four numbers is 301. The numbers are in the ratios $15 : 11 : 8 : 9$. What are the numbers?

Solution 7

Use the properties of multiple ratios:

$$\frac{x}{15} = \frac{y}{11} = \frac{z}{8} = \frac{t}{9} = \frac{x+y+z+t}{15+11+8+9}$$

$$= \frac{301}{43} = 7$$

From here, we can derive each number:

$$x = 15 \times 7 = 105$$

$$y = 11 \times 7 = 77$$

$$z = 8 \times 7 = 56$$

$$t = 9 \times 7 = 63$$

Exercise 8

3300 tons of fuel must be delivered to three neighborhoods. There are 250 inhabitants in one neighborhood, 500 in another, and 350 in the last. If the fuel is divided equally among inhabitants, what quantity of fuel will be delivered to each neighborhood?

Solution 8

The total number of inhabitants is $250 + 500 + 350 = 1100$. Per inhabitant, the consumption is $3300 \div 1100 = 3$ tons. Each neighborhood will receive:

$$250 \times 3 = 750 \text{ tons}$$

$$500 \times 3 = 1500 \text{ tons}$$

$$350 \times 3 = 1050 \text{ tons}$$

Exercise 9

Three positive integers are in the ratios 2 : 4 : 7. The difference between the largest and the smallest numbers is 55. What are the numbers?

Solution 9

Order the numbers and denote them by $a < b < c$. We know that $a + 55 = c$:

$$\frac{a}{2} = \frac{a + 55}{7}$$

Cross-multiply and solve for a:

$$
\begin{aligned}
7a &= 2(a + 55) \\
7a &= 2a + 110 \\
5a &= 110 \\
a &= 22
\end{aligned}
$$

Now it is easy to derive the values of the other two numbers:

$$c = a + 55 = 22 + 55 = 77$$

and

$$
\begin{aligned}
\frac{a}{2} &= \frac{b}{4} \\
4a &= 2b \\
2a &= b \\
b &= 2 \times 22 = 44
\end{aligned}
$$

The numbers are 22, 44, and 77.

Exercise 10

Write proportions using the numbers: 7, 11, 56, and 88.

Solution 10

First, factor the numbers ($56 = 7 \times 8$ and $88 = 11 \times 8$) and identify the two products that are equal:

$$7 \times 88 = 11 \times 56$$

Write all the possible proportions that have the same product of means/extremes:

$$\frac{7}{11} = \frac{88}{56}$$

$$\frac{11}{7} = \frac{56}{88}$$

$$\frac{7}{88} = \frac{11}{56}$$

$$\frac{88}{7} = \frac{56}{11}$$

Exercise 11

Which of the following sets of 4 numbers can form proportions? Check all that apply.

Solution 11

Apply the cross-multiplication property to check whether it is possible to form two equal products:

(A) $\{52, \ 8, \ 13, \ 32\} = \{4 \times 13, \ 8, \ 13, \ 8 \times 4\}$

$52 \times 8 = 13 \times 32$

A family of proportions can be based on the above product.

(B) $\{52, \ 4, \ 13, \ 39\} = \{4 \times 13, \ 4, \ 13, \ 3 \times 13\}$

There is a single factor of 3. We cannot form two equal products with the given numbers.

54

(C) $\{11,\ 88,\ 28,\ 7\} = \{11,\ 11 \times 8,\ 7 \times 4,\ 7\}$

There are 5 factors of 2 ($2^3 = 8$ and $2^2 = 4$). We cannot form two equal products.

(D) $\{2a,\ 5a,\ 10b,\ b\}$

To obtain the factor of 10, $5a$ must multiply $2a$. But $10a^2 \neq 10b^2$. We cannot form a proportion.

(E) If $n \neq m$ there are no possible equal products. We cannot form a proportion.

The answer is (A) only.

Exercise 12

On a map, the distance between two points is represented by an 8 mm long segment. The scale of the map is $\frac{1}{2000000}$. What is the actual distance, in kilometers, between the two points?

Solution 12

$$\frac{1}{2000000} = \frac{8}{x}$$

$$\frac{1}{2 \times 10^6} = \frac{8}{x}$$

$$x = 8 \times 2 \times 10^6 \text{ mm}$$

$$x = 16 \times 10^6 \text{ mm}$$

$$x = 16 \times 10^3 \text{ m}$$

$$x = 16 \text{ km}$$

Exercise 13

A wheel spins 100 times in 2 minutes and 30 seconds. How many times does it spin in 15 minutes?

Solution 13

First, transform quantities to the same unit, respectively. 2 minutes and 30 seconds equals 2.5 minutes. The number of *revolutions* (complete rotations) is proportional to time:

$$\frac{2.5}{15} = \frac{100}{x}$$

$$\frac{25}{150} = \frac{100}{x}$$

$$\frac{1}{6} = \frac{100}{x}$$

$$x = 600$$

The wheel makes 600 revolutions in 15 minutes.

Exercise 14

If the real distance between points A and B is 300 miles, what is the distance in inches that separates them on a map drawn at a scale of 1 : 1500000?

Solution 14

Denote the distance on the map by x. First, convert miles to inches:

$$1 \text{ mile} = 5280 \text{ feet}$$

$$1 \text{ mile} = 63360 \text{ inches}$$

Substitute this value in the proportion:

$$\frac{x}{300 \times 63360} = \frac{1}{1500000}$$

$$\frac{x}{300 \times 63360} = \frac{1}{300 \times 5000}$$

$$\frac{x}{63360} = \frac{1}{5000}$$

$$x \times 5000 = 63660$$

$$x = \frac{63360}{5000}$$

$$x = 12.672 \text{ inches}$$

On the map, the two points are separated by 12.672 inches.

Exercise 15

If x and y are terms in the proportion:

$$\frac{3}{x} = \frac{y}{5}$$

find the value of the expression:

$$E = \frac{xy + 7}{2xy - 8}$$

Solution 15

Multiply means and extremes to find the value of xy:

$$xy = 15$$

Substitute this value in the expression:

$$E = \frac{15 + 7}{2 \times 15 - 8} = \frac{22}{30 - 8} = \frac{22}{22} = 1$$

Notice that, from the information given, it is not possible to compute the values of x and y. It is possible, however, to compute the product of the two and this is all we need.

Exercise 16

Find the value of the ratio $p : q$, knowing that:

$$\frac{8p}{4q - 5p} = 11$$

Solution 16

Strategy 1: Multiply means and extremes and find the ratio:

$$8p = 44q - 55p$$

$$8p + 55p = 44q$$

$$63p = 44q$$

$$\frac{p}{q} = \frac{44}{63}$$

Strategy 2: Divide all the terms by q to single out the ratio. Solve for $p : q$:

$$\frac{\frac{8p}{q}}{4 - \frac{5p}{q}} = 11$$

$$\frac{8\frac{p}{q}}{4 - 5\frac{p}{q}} = 11$$

$$\frac{8x}{4 - 5x} = 11$$

$$8x = 44 - 55x$$

$$63x = 44$$

$$x = \frac{44}{63}$$

Exercise 17

If $p : q$ is the same as $3 : 4$, compute the value of the expression:

$$F = \frac{5q + p}{5q - 3p}$$

Solution 17

Divide both numerator and denominator by q to single out the ratio and substitute its value:

$$
\begin{aligned}
F &= \frac{\frac{5q+p}{q}}{\frac{5q-3p}{q}} \\[2ex]
&= \frac{5 + \frac{p}{q}}{5 - 3 \times \frac{p}{q}} \\[2ex]
&= \frac{5 + \frac{3}{4}}{5 - 3 \times \frac{3}{4}} \\[2ex]
&= \frac{\frac{23}{4}}{\frac{11}{4}} \\[2ex]
&= \frac{23}{11}
\end{aligned}
$$

Exercise 18

Divide the number 625 in three parts that are in the same ratios as the numbers 2, 3, and $3.\overline{3}$.

Solution 18

See the volume "Arithmetic and Number Theory" for practice with non-terminating decimals.

Denote the parts by a, b, and c and use the fact that $a + b + c = 625$.

$$\frac{a}{2} = \frac{b}{3} = \frac{c}{3.\overline{3}} = \frac{a+b+c}{2+3+3.\overline{3}}$$

$$= \frac{625}{8.\overline{3}}$$

$$= \frac{625}{8 + \frac{1}{3}}$$

$$= \frac{625}{\frac{25}{3}}$$

$$= \frac{625 \times 3}{25}$$

$$= \frac{25 \times \cancel{25} \times 3}{\cancel{25}}$$

$$= 75$$

We can now find each part:

$$a = 2 \times 75 = 150$$

$$b = 3 \times 75 = 225$$

$$c = 625 - 150 - 225 = 250$$

Exercise 19

If the time left until the end of the day is $\frac{3}{7}$ of the time that has already passed since the day began, what time is it now?

Solution 19

A day is 24 hours long. Denote the time that has passed since the start of the day by x. This will also be the current time. The time remaining until the end of the day is $24 - x$. We know that:

$$24 - x = \frac{3}{7} \times x$$

This is an equation that we can solve for x:

$$
\begin{aligned}
24 \times 7 - 7x &= 3x \\[2mm]
24 \times 7 &= 10x \\[2mm]
x &= \frac{24 \times 7}{10} \\[2mm]
x &= \frac{12 \times 7}{5} \\[2mm]
x &= \frac{84}{5} \\[2mm]
x &= 16 + \frac{4}{5} \\[2mm]
x &= 16 + \frac{48}{60}
\end{aligned}
$$

The current time is 16:48 in military time notation. In AM-PM notation, this is 4:48 PM.

Exercise 20

A path is 175 feet shorter than another path. One third of the length of one path is equal to three fourths of the length of the other path. Find the lengths of the two paths.

Solution 20

If the length of one path is L then the length of the other path is $L-175$. Using appropriately the longer and the shorter paths, we have:

$$\frac{1}{3} \times L = \frac{3}{4} \times (L - 175)$$

$$\frac{1}{3} \times L = \frac{3}{4} \times L - \frac{3 \times 175}{4}$$

$$\frac{3}{4} \times L - \frac{1}{3} \times L = \frac{3 \times 175}{4}$$

$$(\frac{3}{4} - \frac{1}{3}) \times L = \frac{3 \times 175}{4}$$

$$\frac{9 - 4}{12} \times L = \frac{3 \times 175}{4}$$

$$\frac{5}{3} \times L = 3 \times 175$$

$$L = 9 \times 35 = (10 - 1) \times 35$$

$$L = 350 - 35 = 315 \text{ feet}$$

SOLUTIONS TO PRACTICE TWO

Do not use a calculator for any of the problems!

Exercise 1

A quantity is increased by 10%, then by 15%, and is finally decreased by 20%. Overall, did the quantity increase or decrease? What is the percent increase/decrease?

Solution 1

Denote the initial quantity by Q_{old}. Then Q_{new} is equal to:

$$
\begin{aligned}
Q_{new} &= Q_{old} \times 1.1 \times 1.15 \times 0.8 \\
&= Q_{old} \times 1.012 \\
&= Q_{old} \times \left(1 + \frac{1.2}{100}\right)
\end{aligned}
$$

The quantity increased by 1.2%.

Exercise 2

Two quantities are in a ratio of 5 : 4. If the first quantity increased by 20%, to which of the following did its ratio to the other quantity change?

(A) $3 : 2$

(B) $4 : 5$

(C) $6 : 5$

(D) $1.2 : 5$

Solution 2

Denote the two quantities by Q and P. We know that:

$$
\begin{aligned}
Q : P &= 5 : 4 \\
1.2 \cdot Q : P &= 1.2 \times 5 : 4 \\
1.2 \cdot Q : P &= 6 : 4 \\
1.2 \cdot Q : P &= 3 : 2
\end{aligned}
$$

The correct answer is (A).

Exercise 3

Due to changes in the design of the Mouse Maze, the path to the Cheese has been shortened by 10%. At the same time, a diligent workout routine has increased the Mouse's speed by 25%. By how much percent has the average time of access to the Cheese changed? Is it an increase or a decrease?

Solution 3

Assume the time needed to access the Cheese to be:

$$ t = \frac{D}{v} $$

where D is the distance and v is the speed.

After the changes, the new access time becomes:

$$ T = \frac{0.9 \cdot D}{1.25 \cdot v} $$

We have to figure out how this time is related to the old time:

$$
\begin{aligned}
T &= \frac{0.9 \cdot D}{1.25 \cdot v} \\
&= \frac{90}{125} \times \frac{D}{v} \\
&= \frac{18}{25} \times t \\
&= \frac{72}{100} \times t
\end{aligned}
$$

The new time is 72% of the old time. Therefore, it has decreased by 28%.

Exercise 4

In modern English, 29% of the vocabulary is assumed to be of French origin and 26% of Germanic origin. If the vocabulary grows by 0.5% each decade by adopting words of a different origin, will the difference between the percentage of French-origin words and the percentage of Germanic-origin words be greater or smaller than 3% after 2 decades?

Solution 4

Let us assume that the number of words in today's English vocabulary is T. Then, the number of words of French origin is $0.29 \cdot T$ and the number of words of Germanic origin is $0.26 \cdot T$.

After 2 decades, the size of the vocabulary will be:

$$
T_{\text{new}} = 1.005 \cdot 1.005 \cdot T = 1.010025 \cdot T
$$

The percentage of French words will be:

$$
\frac{0.29 \cdot T}{1.010025 \cdot T} \times 100\% = \frac{29}{101.0025} \times 100\%
$$

The percentage of Germanic words will be:

$$
\frac{0.26 \cdot T}{1.010025 \cdot T} \times 100\% = \frac{26}{101.0025} \times 100\%
$$

The difference between the two percentages will be:

$$\left(\frac{29}{101.0025} - \frac{26}{101.0025}\right) \times 100\% = \frac{29-26}{101.0025} \times 100\%$$
$$= \frac{3}{101.0025} \times 100\%$$
$$= \frac{300}{101.0025}\% < \frac{300}{100}\%$$

This percentage is smaller than 3%.

Exercise 5

The width of a rectangle increased by 60% and its length increased by 20%. By what percentage did its area increase?

Solution 5

The area before the increases is the product of length and width:

$$A_{\text{old}} = W \times L$$

If the width becomes $1.6 \cdot W$ and the length becomes $1.2 \cdot L$, then the new area becomes:

$$A_{\text{new}} = 1.6 \cdot W \times 1.2 \cdot L = 1.92 \cdot W \cdot L = 1.92 \cdot A_{\text{old}}$$

The area increased by 92%.

Exercise 6

The width of a rectangle increased by 30% and its length decreased by 30%. Did its area increase or decrease? By what percent?

Solution 6

The area before the increases is the product of length and width:

$$A_{\text{old}} = W \times L$$

If the width becomes $1.3 \cdot W$ and the length becomes $0.7 \cdot L$, then the new area becomes:

$$A_{\text{new}} = 1.3 \cdot W \times 0.7 \cdot L = 0.91 \cdot W \cdot L = 0.91 \cdot A_{\text{old}}$$

The area decreased by 9%.

Exercise 7

A circle's radius increased by 5%. By what percentage did its area increase?

Solution 7

If the initial radius is R, the area of the circle before the increase was:

$$A_{\text{old}} = \pi R^2$$

The new radius is $1.05 \cdot R$ and the new area is:

$$
\begin{aligned}
A_{\text{new}} &= \pi(1.05 \cdot R)^2 \\
&= \pi \times 1.05^2 \times R^2 \\
&= 1.1025 \times \pi R^2 \\
&= 1.1025 \times A_{\text{old}} \\
&= \left(1 + \frac{10.25}{100}\right) \times A_{\text{old}}
\end{aligned}
$$

The area has increased by 10.25%.

Exercise 8

What percent of N is 10% of 120% of 35% of N?

(A) 3.26%

(B) 4.2%

(C) 40%

(D) 42%

Solution 8

The correct answer choice is (B).

$$
\begin{aligned}
0.1 \times 1.2 \times 0.35 \times N &= 0.042 \times N \\
&= \frac{4.2}{100} \times N \\
&= 4.2\% \times N
\end{aligned}
$$

Exercise 9

If the 30° angle of a triangle increases by 50%, by what percentage does the sum of the other two angles decrease?

Solution 9

The angles of a triangle always add up to 180°. The sum of the other two angles is thus $180° - 30° = 150°$. The 30° angle increases by 50% to 45°. The sum of the other angles decreases to $150° - 15° = 135°$.

$$
\begin{aligned}
135° &= 150° \times \frac{x}{100} \\
x &= \frac{135 \times 100}{150} \\
x &= 90
\end{aligned}
$$

Since 135° is 90% of 150°, the sum of the other two angles decreases by 10%.

Exercise 10

The sum of three numbers is 87. If we increase the first number by 150%, decrease the second one by 25%, and decrease the third one by 5, the results are equal. Find the numbers.

Solution 10

Denote the numbers by a, b, and c. Then, we have:

$$a + b + c = 87$$

as well as:

$$a \times 2.5 = b \times 0.75 = c - 5$$

Express a and c as functions of b:

$$a = b \cdot \frac{75}{250} = b \cdot \frac{3}{10}$$

$$a = 0.3 \cdot b$$

$$c = 0.75 \cdot b + 5$$

and use the sum:

$$
\begin{aligned}
0.3 \cdot b + b + 0.75 \cdot b + 5 &= 87 \\
2.05 \cdot b &= 82 \\
205 \cdot b &= 8200 \\
5 \times 41 \times b &= 2 \times 41 \times 100 \\
\cancel{5} \times \cancel{41} \times b &= 2 \times \cancel{41} \times 20 \times \cancel{5} \\
b &= 2 \times 20 \\
b &= 40
\end{aligned}
$$

Therefore, $b = 40$, $a = 12$, and $c = 35$.

Exercise 11

Alfonso sorted 420 lbs of potatoes in two containers so that 30% of the contents of one container have the same weight as 40% of the contents of the other container. The difference between the amounts of potatoes in the two containers is:

(A) 42 lbs

(B) 50 lbs

(C) 60 lbs

(D) 70 lbs

Solution 11

Denote the two quantities of potatoes by x and y. We know that:

$$x + y = 420$$
$$\frac{30}{100} \cdot x = \frac{40}{100} \cdot y$$

This is a linear system of two equations with two unknowns. From the second equation we infer that $3x = 4y$. We multiply the first equation by 3 to make the computations smoother:

$$3x + 3y = 3 \times 420$$

and we substitute $4y$ for $3x$:

$$
\begin{aligned}
4y + 3y &= 3 \times 420 \\
7y &= 3 \times 420 \\
y &= 3 \times 60 \\
y &= 180 \text{ lbs}
\end{aligned}
$$

From here $x = 240$ lbs and the difference between the two quantitites of potatoes is 60 lbs.

The correct answer is (C).

Exercise 12

What percentage of the positive integers between 34 and 147 are multiples of 8?

Solution 12

There is a multiple of 8 every 8 numbers. There are 4 multiples of 8 between 1 and 34:

$$34 = 4 \times 8 + 2$$

There are 18 multiples of 8 between 1 and 147:

$$147 = 18 \times 8 + 3$$

There are $18 - 4 = 14$ multiples of 8 between 34 and 147. The total number of integers between 34 and 147 is $146 - 34 = 112$. Of the total number of integers in the same interval, the percentage of multiples of 8 is:

$$\frac{14}{146 - 34} \times 100\% = \frac{14}{112} \times 100\% = \frac{1}{8} \times 100\% = 12.5\%$$

Exercise 13

The side of a square increased by 50%. By what percentage did its area increase?

Solution 13

Denote the side by s. The initial area was s^2.

After increasing, the side became equal to $s \times 1.5$. The increased area was:

$$A_{\text{new}} = (s \times 1.5)^2 = s^2 \times 2.25 = A_{\text{old}} \times 2.25$$

The percent increase of the area is:

$$\frac{A_{\text{new}} - A_{\text{old}}}{A_{\text{old}}} \times 100\% \;=\; \frac{A_{\text{old}} \times 2.25 - A_{\text{old}}}{A_{\text{old}}} \times 100\%$$

$$=\; \frac{A_{\text{old}} \cdot (2.25 - 1)}{A_{\text{old}}} \times 100\%$$

$$=\; 1.25 \times 100\%$$

$$=\; 125\%$$

Exercise 14

Due to drought conditions, only a third of Dina's tomato plants survived last week's heat. What was the percent decrease in tomato plants?

Solution 14

The new and the old populations are related as in:

$$N_{\text{new}} = \frac{1}{3} \cdot N_{\text{old}}$$

The percent decrease is $66.\overline{6}\%$:

$$\frac{N_{\text{new}} - N_{\text{old}}}{N_{\text{old}}} \times 100\% \;=\; \frac{N_{\text{new}} - 3N_{\text{new}}}{3N_{\text{new}}} \times 100\%$$

$$=\; \frac{N_{\text{new}}(1 - 3)}{3N_{\text{new}}} \times 100\%$$

$$=\; \frac{-2}{3} \times 100\%$$

$$=\; -\frac{200}{3}\%$$

$$=\; -66.\overline{6}\%$$

Exercise 15

A store uses a profit/loss simulator to determine the effect of multiple rebates in prices. The store manager inputs the products he wants to put on sale and applies a rebate of 15%. The simulator shows a profit of 25% for that set of products. The manager applies another rebate and the profit goes down to 0%. Find the value of the second rebate.

Solution 15

Denote the cost of the products to the store by C and the selling price by S.

Assume the profit to be the difference between the selling price and the cost: $S - C$.

After the rebate, the selling price becomes $0.85 \times S$. Selling at this price still brings in 25% more than the cost of buying the products:

$$0.85 \times S = 1.25 \times C$$

Find the factor that, multiplied by the rebated selling price, makes the profit zero. When the profit is zero, the selling price is exactly equal to the cost of buying the products:

$$x \times 0.85 \times S = C$$

From the previous equation, we find that:

$$x = \frac{1}{1.25} = 0.8 = 80\%$$

The second rebate must be equal to 20%.

SOLUTIONS TO PRACTICE THREE

Do not use a calculator for any of the problems!

Exercise 1

When we add 100 grams of water to a water and salt solution that has a salt concentration of 20%, the salt concentration drops to 16%. What is the final mass of the solution?

Solution 1

Write an equation that sets the amounts of pure salt before and after the mixing to be equal. Before the mixing, the mass in grams of the solution was M. After mixing, the mass of the solution became $M+100$ since 100 grams of water were added.

$$\frac{20}{100} \times M = \frac{16}{100} \times (M + 100)$$

Now solve for the initial mass of the solution:

$$
\begin{aligned}
20 \cdot M &= 16 \times (M + 100) \\
20 \cdot M &= 16 \cdot M + 1600 \\
4 \cdot M &= 1600 \\
M &= 400 \text{ g}
\end{aligned}
$$

The final mass of the solution is 500 grams.

Exercise 2

Stephan had a box of tennis balls. 80% of the balls were yellow and the rest were green. What percentage of the green balls must be replaced by yellow balls so that the percentage of yellow balls in the box will become 85%?

Solution 2

Strategy 1:

This is a more formal solution that can be reused in complicated problems.

Assume the total number of balls is N. The number of yellow balls is $\frac{80}{100} \times N$ and the number of green balls is $\frac{20}{100} \times N$. The number of green balls that are replaced by yellow balls is $\frac{20}{100} \times \frac{x}{100} \times N$. The total number of yellow balls:

$$\frac{80}{100} \times N + \frac{20}{100} \times \frac{x}{100} \times N = \frac{85}{100} \times N$$

$$8000 \cdot N + 20 \cdot x \cdot N = 8500 \cdot N$$

$$20 \cdot x = 500$$

$$x = 25$$

25% of the green balls must be replaced by yellow balls.

Strategy 2:

Assume a total number of 100 balls, of which 80 are yellow and 20 are green. To have 85 yellow balls, 5 green balls must be replaced by yellow balls. This is 25% of the number of green balls.

Exercise 3

Dina and Lila made a fruit punch that is 50% juice and 50% water. They drank half of the punch and refilled the jug with water. Then, they drank half of the punch again and refilled the jug with juice. What was the final concentration of the juice?

Solution 3

Strategy 1:

After the first refill, the mixture was 75% water and 25% juice. After they drank half of the contents of the jug again, the amount of juice in the remaining liquid equaled 12.5% of the capacity of the jug. After refilling the jug with juice, the juice represented 62.5% of the capacity of the jug.

Strategy 2:

Assume the jug contains 1000 mL, of which 500 mL are juice and 500 mL are water.

After they drank half of the mixture, there were 250 mL of juice and 250 mL of water left. After they refilled the jug with water, there were 250 mL of juice and 750 mL of water in the mixture.

After they drank half of the mixture again, there were 125 mL of juice and 375 mL of water in the mixture. After a refill with juice, there were 625 mL of juice and 375 mL of water in the mixture.

The final concentration of the juice was:

$$\frac{625}{1000} = \frac{62.5}{100} = 62.5\%$$

Exercise 4

Ali and Baba have a fortune in gold and silver coins. If they spent 600 gold coins and 200 silver coins, the remaining gold coins would represent 55% of the total number of coins. If they spent 200 gold coins and 400 silver coins, the remaining gold coins would represent 65% of the total number of coins. How many coins do Ali and Baba have?

Solution 4

Denote the number of gold coins with G and the total number of coins with N.

In the first hypothesis, the total number of coins is reduced by 800 while the number of gold coins is reduced by 600:

$$G - 600 = \frac{55}{100} \times (N - 800) = \frac{11}{20} \times (N - 800)$$

In the second hypothesis, the total number of coins is reduced by 600 while the number of gold coins is reduced by 200:

$$G - 200 = \frac{65}{100} \times (N - 600) = \frac{13}{20} \times (N - 600)$$

Subtract the first equation from the second one:

$$G - 200 - (G - 600) = \frac{13}{20} \times (N - 600) - \frac{11}{20} \times (N - 800)$$

$$G - 200 - G + 600 = \frac{13}{20} \times (N - 600) - \frac{11}{20} \times (N - 800)$$

$$400 \times 20 = 13 \cdot N - 13 \times 600 - 11 \cdot N + 11 \times 800$$

$$8000 = 2 \cdot N + 1000$$

$$2 \cdot N = 7000$$

$$N = 3500$$

The treasure consists of 3500 coins.

Exercise 5

A mixture of two copper ores, one that is 35% copper and another that is 42% copper, has a copper concentration of 40%. If we add 350 kilograms of the less concentrated ore to the mixture, the copper concentration in the mixture drops to 38%. What is the total mass, in kilograms, of the original mixture?

Solution 5

Denote the mass of the less concentrated ore by A and the mass of the more concentrated ore by B. The total mass of the mixture is $A + B$. Then, before adding more ore:

$$A \times \frac{35}{100} + B \times \frac{42}{100} = (A + B) \times \frac{40}{100}$$

After adding more ore:

$$(A + 350) \times \frac{35}{100} + B \times \frac{42}{100} = (A + B + 350) \times \frac{38}{100}$$

Both equations are in kilograms.

This is a linear system of two equations with two unknowns.

$$35 \cdot A + 42 \cdot B + 350 \times 35 \;=\; 38 \cdot A + 38 \cdot B + 38 \times 350$$

$$35 \cdot A + 42 \cdot B \;=\; 40 \cdot A + 40 \cdot B$$

Since we only need the total mass $A + B$, it is advantageous to subtract the equations:

$$350 \times 35 \;=\; -2 \cdot A - 2 \cdot B + 38 \times 350$$

$$2(A + B) \;=\; 350 \times (38 - 35)$$

$$A + B \;=\; 175 \times 5$$

$$A + B \;=\; 525 \text{ kg}$$

Initially, the mixture contained 150 kg of the less concentrated ore and 375 kg of the more concentrated ore. In total, the mixture weighed $150 + 375 = 525$ kg.

Exercise 6

A two liter solution consists of 95% water and 5% vinegar. How much water should we add to it so that the vinegar concentration falls to 2%?

Solution 6

Denote the amount of added water by W, in mL. The solution has 2000 mL now and will have $2000 + W$ mL after we add the extra water.

The amount of pure vinegar does not change. We can write an equation that sets the amount of vinegar before the change equal to the amount of vinegar after the change:

$$2000 \times \frac{5}{100} = (2000 + W) \times \frac{2}{100}$$

and solve it for W:

$$
\begin{aligned}
2000 \times 5 &= (2000 + W) \times 2 \\
10000 &= 4000 + 2 \cdot W \\
6000 &= 2 \cdot W \\
W &= 3000 \text{ mL}
\end{aligned}
$$

We have to add 3 L of water.

Exercise 7

Dina dissolved some sugar in water so that the solution was 12% sugar by mass. Lila did the same but her solution was 18% sugar by mass. They poured both solutions into a pitcher and added another 150 g of water. The pitcher now contained 1200 g of solution with a sugar concentration of 15%. How many grams of solution did each of them have initially?

Solution 7

Denote the volume of Dina's solution by D and the volume of Lila's solution by L. The total amount of sugar is the same before and after the mixing.

$$D \times \frac{12}{100} + L \times \frac{18}{100} = (D + L + 150) \times \frac{15}{100}$$

We also know that $D + L + 150 = 1200$. This means we have a linear system of two equations with two unknowns.

$$
\begin{aligned}
D + L &= 1200 - 150 \\
D + L &= 1050 \\
D &= 1050 - L
\end{aligned}
$$

We can substitute this in the first equation:

$$
\begin{aligned}
12 \cdot D + 18 \cdot L &= 1200 \times 15 \\
4 \cdot D + 6 \cdot L &= 1200 \times 5 \\
2 \cdot D + 3 \cdot L &= 600 \times 5 \\
2 \times (1050 - L) + 3 \cdot L &= 3000 \\
2100 + L &= 3000 \\
L &= 900 \text{ g}
\end{aligned}
$$

Lila had 900 grams of solution and Dina had 150 grams of solution.

Exercise 8

Amira has a bag of blocks. 60% of them are green and the rest are blue. If Amira paints 30% of the green blocks blue and 40% of the blue blocks green, what percentage of the blocks will be green?

Solution 8

Strategy 1:

Denote the total number of blocks by N. The number of green blocks is $\frac{6}{10}N$ and the number of blue blocks is $\frac{4}{10}N$. If Amira paints 40% of the blue blocks green and 30% of the green blocks blue, then the number of green blocks becomes:

$$= \frac{6}{10} \times N - \frac{6}{10} \times \frac{3}{10}N + \frac{4}{10} \times \frac{4}{10} \times N$$
$$= N \times \left(\frac{60}{100} - \frac{18}{100} + \frac{16}{100} \right)$$
$$= N \times \frac{58}{100}$$

58% of the blocks would be green.

Strategy 2:

Assume the total number of blocks to be 100. Amira has 60 green blocks and 40 blue blocks to start with. She paints 16 blue blocks in green and 18 green blocks blue. At the end of the painting, there are $60 - 18 + 16 = 58$ green blocks.

58% of the blocks would be green.

Exercise 9

How many mL of pure alcohol do we have to mix with a 28% concentrated solution of alcohol in water in order to obtain 360 mL of solution that is 35% concentrated?

Solution 9

Denote the number of mL of pure alcohol by x. The volume of the 28% concentrated solution is $360 - x$ mL. x plus the amount of alcohol in the first mixture equals the amount of alcohol in the final mixture:

$$x + (360 - x) \times \frac{28}{100} = 360 \times \frac{35}{100}$$

$$100 \times x + (360 - x) \times 28 = 360 \times 35$$

$$100 \times x + 360 \times 28 - 28 \times x = 360 \times 35$$

$$72 \times x = 360 \times (35 - 28)$$

$$72 \times x = 360 \times 7$$

$$9 \times 8 \times x = 9 \times 40 \times 7$$

$$\cancel{9} \times 8 \times x = \cancel{9} \times 40 \times 7$$

$$x = 35 \text{ mL}$$

We need 35 mL of pure alcohol.

Exercise 10

We have 200 mL of pure alcohol and 1200 mL of 15% concentrated alcohol in water. What is the largest amount, in mL, of 40% concentrated solution of alcohol in water we can make?

Solution 10

Assume we use x mL of the 15% concentrated solution and all the pure alcohol that is available:

$$200 + x \times \frac{15}{100} = (200 + x) \times \frac{40}{100}$$

$$200 \times 100 + 15 \times x = 200 \times 40 + 40 \times x$$

$$200 \times 60 = 25 \times x$$

$$8 \times 60 = x$$

$$x = 480 \text{ mL}$$

It is possible to make $480 + 200 = 680$ mL of 40% concentrated alcohol in water.

Solutions to Practice Four

Do not use a calculator for any of the problems!

Exercise 1

A team of 3 masons can finish a job in 5 days if they work 8 hours per day. How many days will it take 4 masons, working only 6 hours per day, to complete the same job? (Assume all the masons work at the same rate.)

Solution 1

Use the method of reduction to unity.

masons	days	hours	to complete the job
3	5	8	3 masons work 5 days for 8 hours each day
1	15	8	1 mason works 15 days for 8 hours each day
1	120	1	1 mason works 120 days for 1 hour each day
4	30	1	4 masons work 30 days for 1 hour each day
4	5	6	4 masons work 5 days for 6 hours each day

It will take 5 days.

Exercise 2

To empty a pool, we can use 3 taps. If the first tap is open for 2 hours, the second tap is open for 3 hours, and the third tap is open for 6 hours, 22000 liters of water are going to flow out. If we run the first tap for 3 hours, the second tap for 2 hours, and the third tap for 6 hours, 21000 liters of water will flow out. If the first and the second taps are run for 2 hours and the third tap for 3 hours, 14500 liters of water flow out. How many liters flow through each of the taps in one hour?

Solution 2

Denote the number of liters that flow per hour through each tap, respectively, by a, b, and c. Then we have:

$$2a + 3b + 6c = 22000$$

$$3a + 2b + 6c = 21000$$

$$2a + 2b + 3c = 14500$$

We can solve this system of linear equations to find the values of a, b, and c.

Subtract the second equation from the first:

$$b - a = 1000$$

$$b = a + 1000$$

Multiply the third equation by 2 and subtract the first equation from it:

$$4a + 4b + 6c = 29000$$

$$2a + 3b + 6c = 22000$$

$$2a + b = 7000$$

Substitute $b = a + 1000$:

$$2a + a + 1000 = 7000$$

$$3a = 6000$$

$$a = 2000 \text{ L/hr}$$

Then $b = 3000$ L/hr and we can obtain c from any of the equations:

$$2a + 2b + 3c = 14500$$

$$2 \cdot 2000 + 2 \cdot 3000 + 3c = 14500$$

$$10000 + 3c = 14500$$

$$3c = 4500$$

$$c = 1500 \text{ L/hr}$$

Exercise 3

A crew of workers is using two types of hoses to fill a container. Water flows through one type of hose at a rate of 250 liters/hour and through the other type at a rate of 270 liters/hour. In an hour, 1060 liters of water flow into the container. How many hoses of each type are there?

Solution 3

Denote the numbers of hoses by n and m. The amount of water that flows hourly through these hoses is $250 \cdot n + 270 \cdot m$ and must equal 1060 liters. Since both n and m must be integer, this is a linear Diophantine equation (see the volume "Arithmetic and Number Theory") with 2 unknowns:

$$250 \cdot n + 270 \cdot m = 1060$$

$$25 \cdot n + 27 \cdot m = 106$$

To obtain a last digit of 6, let us consider that multiples of 5 always end in 5 or 0. Therefore, $27m$ must end in 1 or 6. Indeed:

$$25 + 27 \times 3 = 106$$

There is 1 hose of the first type and 3 hoses of the second type.

Exercise 4

A railroad is due for maintenance. If 12 workers can repair half the length in 28 days, how many days will it take to repair the other half if 4 workers quit?

Solution 4

The number of workers and the time it takes to complete the job vary inversely. Therefore:

$$12 \times 28 = 8 \times D$$

We solve for the number of days D:

$$D = \frac{12 \times 28}{8} = \frac{6 \times \cancel{2} \times \cancel{4} \times 7}{\cancel{8}} = 6 \times 7 = 42 \text{ days}$$

Exercise 5

A 10 person team completes a task in 18 days. Another 18 people team completes the same task in 12 days. How long will it take a team formed of 5 people from the first group and 12 people from the second group to finish the job? (Assume that the workers in each group work at the same rate, and that a worker in group 1 works at a different rate from a worker in group 2.)

Solution 5

Use reduction to unity to find out what part of the job each worker completes in 1 day.

For a member of the first team:

workers	days	to complete the job	
10	18	10	workers take 18 days
1	180	1	worker takes 180 days
1	1	$\dfrac{1}{180}$	
5	1	$\dfrac{5}{180}$	5 workers in 1 day

For a member of the second team:

workers	days	to complete the job	
18	12	18	workers take 12 days
1	216	1	worker takes 216 days
1	1	$\dfrac{1}{216}$	
12	1	$\dfrac{12}{216}$	12 workers in 1 day

5 workers from the first group and 12 workers from the second group complete daily:

$$\frac{5}{180} + \frac{12}{216} = \frac{1}{36} + \frac{1}{18} = \frac{3}{36} = \frac{1}{12}$$

of the job.

It will take the new team 12 days to complete the job.

Exercise 6

Two faucets can fill a tub if they run simultaneously for 12 hours. After 8 hours of simultaneous operation, one of the faucets is shut. It took the other faucet 10 hours to fill the tub completely. What time would it take each faucet to fill the empty tub alone?

Solution 6

Denote the flow rate of one faucet by a and of the other by b. Then, the capacity (volume) of the full tub is $12a + 12b$, as well as $8a + 8b + 10b$. These two quantities must be equal, therefore:

$$
\begin{aligned}
12a + 12b &= 8a + 8b + 10b \\
12a + 12b &= 8a + 18b \\
4a &= 6b \\
2a &= 3b
\end{aligned}
$$

The full tub has a capacity of $12a + 12b$, which can also be written as:

$$
\begin{aligned}
12a + 12b &= 12a + 4 \times 3b \\
&= 12a + 4 \times 2a \\
&= 12a + 8a = 20a
\end{aligned}
$$

which means it takes 20 hours for the faucet with flow rate a to fill the tub completely.

Similarly,

$$
\begin{aligned}
12a + 12b &= 6 \times 2a + 12b \\
&= 6 \times 3b + 12b \\
&= 18b + 12b = 30b
\end{aligned}
$$

which means it takes 30 hours for the faucet with flow rate b to fill the tub completely.

Exercise 7

Dina and Amira's families rented a house to go on vacation. They paid 1400 dollars in total. Amira's family has 3 members and stayed for 5 days. Dina's family has 5 members and stayed for 4 days. How should the two families share the expenses fairly?

Solution 7

Let us calculate a price per person per day. If the residence were used by only one person per day, Amira's family would use it for $3 \times 5 = 15$ days, while Dina's family would use it for $5 \times 4 = 20$ days. In total, the residence would be used for 45 days. Each day would cost $1400 \div 35 = 40$ dollars. Therefore, the price per person per day is 40 dollars.

Amira's family should pay $15 \times 40 = 600$ dollars.

Dina's family should pay $20 \times 40 = 800$ dollars.

Exercise 8

Dina and Lila have some old watches that do not work properly anymore. Dina's watch advances by 4 minutes every hour and Lila's watch falls behind by 6 minutes every hour. If they both set their watches to show the same time in the morning, by evening Dina's watch shows 7:36 PM while Lila's watch shows 6:06 PM. At what time did they set their watches to show the same time?

Solution 8

If started at the same time, the two watches would end up 10 minutes apart after the first hour. In the evening they are 1 hour and 30 minutes apart (90 minutes) which means they have been functioning for 9 hours. They were set to show the same time at $10 : 00$ AM.

Exercise 9

Ali and his donkey start walking from Ali's house towards the Date Palm Oasis. Ali walks three times faster than the donkey. He reaches the oasis, fills a bag with dates and starts back towards home. After leaving the oasis, Ali meets his donkey one third of the way home and decides to ride him all the way home. If it took Ali 15 minutes to fill the bag, how long will it take them to get home from the meeting point?

Solution 9

Let us make some notations first:

- D is the distance from Ali's home to the Date Palm Oasis;

- q is the donkey's walking speed and $3q$ is Ali's walking speed, both in miles per hour.

The distance walked by Ali until they meet is $D + \frac{1}{3}D = \frac{4}{3}D$. The distance walked by the donkey is $\frac{2}{3}D$. Make an equation for the time it takes them to arrive at the meeting point. Remember, it takes Ali 15 minutes ($\frac{1}{4}$ hrs) to fill the bag with dates.

$$\frac{\frac{4 \cdot D}{3}}{3q} + \frac{1}{4} = \frac{\frac{2 \cdot D}{3}}{q}$$

$$\frac{4}{3} \times \frac{D}{3q} + \frac{1}{4} = \frac{2}{3} \times \frac{D}{q}$$

$$\frac{1}{4} = \frac{D}{q} \times \left(\frac{2}{3} - \frac{4}{9}\right)$$

$$\frac{1}{4} = \frac{D}{q} \times \frac{2}{9}$$

$$\frac{D}{q} = \frac{1}{4} \times \frac{9}{2}$$

$$\frac{D}{q} = \frac{9}{8}$$

We have found the time it takes the donkey to walk the distance D. We need the time it takes the donkey to walk from the meeting point

to Ali's house:
$$t = \frac{D}{q} \times \frac{2}{3} = \frac{9}{8} \times \frac{2}{3} = \frac{3}{4} \text{ hrs}$$

It takes them 45 minutes to walk home from the meeting point.

Notice that it is not possible to determine the distance or their walking speeds.

Exercise 10

Dina and Lila went hiking in the park with their mother. They were driving home when their mother had to stop for a train. Between the time the locomotive passed them and the time the last car passed them, they waited at the barrier for 1.5 minutes. Farther on, the train had to pass through a 1.5 mile long tunnel which it did in 6 minutes. If the speed of the train was constant, what was the length of the train?

Solution 10

This is a classic problem in which the moving objects are not pointlike. The train's length plays a role. Let us denote the length of the train with L.

There are two situations in the problem. In both of them, the speed of the train is the same. We will use this information to write up an equation between the speed of the train in the first and second situations:

$$\frac{L}{1.5} = \frac{L + 1.5}{6}$$

$$\frac{2L}{3} = \frac{2L + 3}{12}$$

$$24L = 6L + 9$$

$$18L = 9$$

$$L = \frac{9}{18} = 0.5 \text{ miles}$$

The train was half a mile long.

Exercise 11

The little train at the Zoo runs in a loop that is 40 times its length. Assuming the train never stops and runs at a constant speed, what percentage of the time is the train directly in front of the cashier's booth?

Solution 11

Another problem with vehicles that are not pointlike! Denote the speed of the train by s and the length of the train by L. The time it takes the train to pass the cashier's booth is:

$$\frac{L}{s}$$

The time it takes the train to complete a loop is:

$$\frac{40L}{s}$$

The percentage is:

$$
\begin{aligned}
p &= \frac{L}{s} \div \frac{40L}{s} \times 100\% \\[2mm]
&= \frac{L}{s} \times \frac{s}{40L} \times 100\% \\[2mm]
&= \frac{1}{40} \times 100\% \\[2mm]
&= 0.025 \times 100\% \\[2mm]
&= 2.5\%
\end{aligned}
$$

Exercise 12

Stephan is a member of a sailing team. In a race, they had to sail from buoy A to B to C to D, and then sail back on the same route. They sailed from A to B at a speed of 9 knots, from B to C at a speed of 4 knots, and from C to D at a speed of 3 knots. On the way back, the speeds were exactly reversed. The way from A to D took twice as long as the way from D to A. Which of the following could be the ratio AB:BC:CD?

(A) $1 : 4 : 14$

(B) $9 : 4 : 3$

(C) $3 : 4 : 9$

(D) $1 : 7 : 13$

Solution 12

Let us denote the distances as follows: $AB = a$, $BC = b$, and $CD = c$. Then, the times for sailing from A to D and from D to A are related by the equation:

$$\frac{a}{9} + \frac{b}{4} + \frac{c}{3} = 2 \left(\frac{a}{3} + \frac{b}{4} + \frac{c}{9} \right)$$

$$\frac{c}{3} - \frac{2c}{9} = \frac{2a}{3} - \frac{a}{9} + \frac{b}{4}$$

$$\frac{c}{9} = \frac{5a}{9} + \frac{b}{4}$$

$$4c = 20a + 9b$$

The ratio $a : b : c = 1 : 4 : 14$ satisfies the equation. The correct answer choice is (A).

Exercise 13

A vehicle's speed changes as follows in a 2 hour time interval. What is the distance covered by the vehicle during this time?

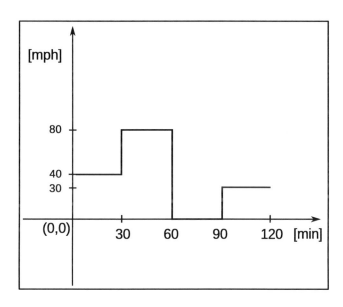

Solution 13

Notice that the distance covered is equal to the sum of the areas of the rectangles. Express the time in hours to obtain the distance in miles:

$$40 \times 0.5 + 80 \times 0.5 + 30 \times 0.5 = 20 + 40 + 15 = 75 \text{ miles}$$

Exercise 14

A vehicle's speed changes as follows in a 2 hour time interval. What is the distance covered by the vehicle during this time?

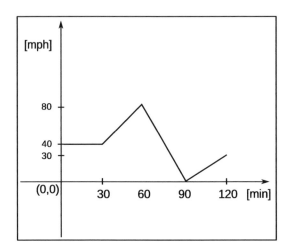

Solution 14

This problem appears to be more complicated since the speed is not constant all the time. However, remember the observation we made in the previous problem, that *the distance covered is equal to the area below the graph.*

Let us divide the area into simpler figures:

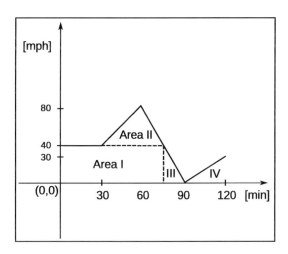

The area I is: $40 \times 1.25 = 50$ miles.

The area II is a triangle with base 45 min = 0.75 hrs and height 40 mph. The distance is:

$$D_{\text{II}} = \frac{40 \times 0.75}{2} = 15 \text{ miles}$$

The area III is a triangle with base 15 minutes and height 40 mph. The distance covered is:

$$D_{\text{III}} = \frac{0.25 \times 40}{2} = 5 \text{ miles}$$

The area IV is a triangle with base 30 minutes and height 30 mph. The distance covered is:

$$D_{\text{IV}} = \frac{0.5 \times 30}{2} = 7.5 \text{ miles}$$

The total distance covered is $50 + 15 + 5 + 7.5 = 77.5$ miles.

Exercise 15

Stephan had to travel to a tennis tournament. He traveled by plane for 4 hours at an average speed of 600 mph and then by car for 2 hours at an average speed of 75 mph. What was his average speed overall?

Solution 15

To compute the average speed we have to compute the total distance and divide it by the total time. The total distance is $600 \times 4 = 2400$ miles by plane plus $2 \times 75 = 150$ miles by car. The average speed is:

$$v = \frac{2400 + 150}{6} = \frac{2550}{6} = 425 \text{ mph}$$

Exercise 16

A mouse positioned at corner A of a square runs along the perimeter of the square on the shortest path towards point P, the midpoint of BC. A snail positioned in the center of the square at point O crawls towards P on the shortest possible path. If they leave at the same time and arrive at P simultaneously, what is the ratio between the speed of the snail and the speed of the mouse?

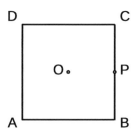

(A) $3 : 1$

(B) $2 : 1$

(C) $1 : 2$

(D) $1 : 3$

Solution 16

Denote the length of the side of the square by $2a$. Then, $OP = a$ since it is half the length of the side. The mouse covers $2a + a = 3a$ units in the same time as the snail covers a units. The ratio is $1 : 3$. The correct answer choice is (D).

Exercise 17

Dina and Lila wanted to go to a Hallowe'en party but they only had one broom. So they decided that Dina could use the broom while Lila walked and that Dina would leave the broom somewhere along the way for Lila to find and use to get to their destination. Both of them walked at 2 miles per hour and rode the broom at 5 miles per hour. If Lila arrived 36 minutes before Dina and the broom remained unused for 18 minutes, how far from the party is their house?

Solution 17

Let use denote the point where Dina left the broom by M and the distance from their house to M by x. It took Dina $\dfrac{x}{5}$ hours to get from home to M and it took Lila $\dfrac{x}{2}$ hours to get from home to M. The difference between these times represents the time the broom remained unused:

$$\frac{x}{2} - \frac{x}{5} = \frac{18}{60}$$

$$\frac{5x - 2x}{10} = \frac{18}{60}$$

$$3x = \frac{180}{60}$$

$$x = 1 \text{ mile}$$

We now know that the broom was left 1 mile away from home.

Let us denote the distance from their home to the party by D. Then, from M to the party, the distance is $D - 1$ miles. The time it took Dina to travel from home to the party, in hours, was:

$$\frac{1}{5} + \frac{D-1}{2}$$

The time it took Lila to travel from home to the party, in hours, was:

$$\frac{1}{2} + \frac{D-1}{5}$$

Dina's time was 36 minutes longer:

$$\frac{1}{5} + \frac{D-1}{2} = \frac{1}{2} + \frac{D-1}{5} + \frac{36}{60}$$

$$\frac{2}{10} + \frac{5D-5}{10} = \frac{5}{10} + \frac{2D-2}{10} + \frac{6}{10}$$

$$2 + 5D - 5 = 5 + 2D - 2 + 6$$

$$5D - 3 = 2D + 9$$

$$3D = 12$$

$$D = 4 \text{ miles}$$

Solutions to Miscellaneous Practice

Do not use a calculator for any of the problems!

Exercise 1

If the length of a rectangle is increased by 0.5 cm, its area increases by 2.5 cm^2. If the width of the same rectangle is increased by 0.25 cm, its area increases by 1.25 cm^2. Which of the following is true?

(A) The rectangle's width is double its length.

(B) The rectangle's length is double its width.

(C) The rectangle's length is equal to its width.

(D) The rectangle's length is five times its width.

Solution 1 Denote the dimensions of the initial rectangle by W and L. Its area will be $W \times L$.

If the length becomes $L_{\text{new}} = L + 0.5$ the area becomes:

$$A_{\text{new}} = W \cdot (L + 0.5) = W \cdot L + 0.5 \cdot W = A_{\text{old}} + 0.5 \cdot W$$

Knowing the difference from the old area, we can calculate W:

$$
\begin{aligned}
0.5 \cdot W &= 2.5 \\
W &= 25 \div 5 \\
W &= 5 \text{ cm}
\end{aligned}
$$

101

We can calculate L in a similar manner. If the width becomes $W_{new} = W + 0.25$ then the new area:

$$A_{new} = (W + 0.25) \cdot L = W \cdot L + 0.25 \cdot L = A_{old} + 0.25 \cdot L$$

where we know the difference between the old and new areas:

$$
\begin{aligned}
0.25 \cdot L &= 1.25 \\
L &= 1.25 \div 0.25 \\
L &= 125 \div 25 \\
L &= 5 \text{ cm}
\end{aligned}
$$

The width and the length are equal. Therefore, the rectangle is a square. The correct answer is (C).

Exercise 2

A metal container is filled with liquid to $x\%$ of its total volume. The container is heated. As a result of thermal expansion, the volume of the container increases by 8% and the volume of the liquid increases by 20%. What must x be if the container is now completely full?

Solution 2

The new volume of the container is:

$$V_{new} = V + V \times \frac{8}{100} = 1.08 \cdot V$$

The new volume of the liquid is:

$$L_{new} = L + L \times \frac{20}{100} = 1.2 \cdot L$$

These two volumes must be equal:

$$1.08 \cdot V = 1.2 \cdot L$$

Since:

$$L = V \times \frac{x}{100}$$

we can substitute L in the previous equation and solve for x:

$$
\begin{aligned}
1.08 \cdot V &= 1.2 \cdot V \times \frac{x}{100} \\
108 \cdot V &= 1.2 \cdot V \cdot x \\
108 &= 1.2x \\
x &= 108 \div 1.2 \\
x &= 1080 \div 12 \\
x &= 90
\end{aligned}
$$

Initially, the container was filled to 90% of its capacity.

Exercise 3

96% of the volume of a cube is filled with water. If all the sides of the cube double in length, what percentage of the new volume will be filled with water? (Assume the amount of water does not change.)

Solution 3

If the sides double in length, the volume becomes 8 times larger:

$$
\begin{aligned}
S &= 2s \\
S^3 &= (2s)^3 \\
S^3 &= 8s^3 \\
V &= 8v
\end{aligned}
$$

The volume of water is:

$$
w = v \times \frac{96}{100}
$$

The water represents x percent of the new volume:

$$w = V \times \frac{x}{100}$$

$$v \times \frac{96}{100} = 8v \times \frac{x}{100}$$

$$96 = 8x$$

$$x = 12$$

12% of the new volume is filled with water.

Exercise 4

A flock of ducklings were crossing a river. 10 of them swam to the other side of the river and half of the remaining ducklings followed them. At that point, there were one third as many ducklings that had not crossed as there were ducklings that crossed. How many ducklings were there in total?

Solution 4

Strategy 1: Denote the total number of ducklings by D. 10 of them swam to the other side and were followed by $\dfrac{D-10}{2}$ ducklings. At this point, there were $\dfrac{D-10}{2}$ ducklings that had not crossed, which represent one third of the number that crossed the river:

$$\frac{D-10}{2} = \frac{1}{3} \times \left(10 + \frac{D-10}{2}\right)$$

$$3 \times \frac{D-10}{2} = 10 + \frac{D-10}{2}$$

$$\cancel{2} \times \frac{D-10}{\cancel{2}} = 10$$

$$D - 10 = 10$$

$$D = 20$$

There were 20 ducklings in total.

Strategy 2: Denote the ducklings that remained on the home side by d. Then, the number of ducklings that crossed is $3d$ and it consists of 10 ducklings plus d ducklings. Therefore:

$$3d = d + 10$$
$$d = 5$$

There must be 15 ducklings on the other side, to a total of 20 ducklings.

Exercise 5

In two containers, one glass and one plastic, there are 26 liters of grape juice. If we transfer 3 liters from the glass container to the plastic container, there are 8 liters less in the glass container than in the plastic container. How many liters of juice were there in the glass container to start with?

Solution 5

Initially, there were G and P liters of juice, respectively. After the transfer, the quantities became: $G - 3$ and $P + 3$. We know that:

$$G - 3 = P + 3 - 8$$
$$G = P + 6 - 8$$
$$G = P - 2$$

We can now use the total amount of juice:

$$26 = G + P$$
$$26 = P - 2 + P$$
$$26 + 2 = 2P$$
$$28 = 2P$$
$$P = 14 \text{ L}$$

There were 14 liters in the plastic container and 12 liters in the glass container.

Exercise 6

Stephan was driving to a tennis tournament. After he drove $\frac{2}{7}$ of the total route, he found that were 150 miles left to cover before getting halfway to destination. How long was Stephan's route?

Solution 6

Denote the total distance with D and set up the equation:

$$\frac{2}{7} \times D + 150 = \frac{D}{2}$$

Solve for D:

$$\frac{D}{2} - \frac{2}{7} \times D = 150$$

$$\frac{7 \cdot D}{14} - \frac{4 \cdot D}{14} = 150$$

$$\frac{3 \cdot D}{14} = 150$$

$$D = \frac{150 \times 14}{3}$$

$$D = 50 \times 14$$

$$D = 700 \text{ miles}$$

Exercise 7

The numbers a, b, and c are in the ratios $4 : 7 : 9$. Compute the ratio:

$$T = \frac{a + 3b}{b + 2c}$$

Solution 7

$$\frac{a}{4} = \frac{b}{7} = \frac{c}{9}$$

Divide the numerator and the denominator of the expression by b:

$$T = \frac{\frac{a}{b} + 3}{1 + 2 \times \frac{c}{b}}$$

and compute the required ratios:

$$\frac{a}{b} = \frac{4}{7}$$

$$\frac{c}{b} = \frac{9}{7}$$

Now substitute the values of the ratios in the expression:

$$T = \frac{\frac{4}{7} + 3}{1 + 2 \times \frac{9}{7}}$$

$$= \frac{\frac{4+21}{7}}{\frac{7+18}{7}}$$

$$= \frac{25}{7} \div \frac{25}{7}$$

$$= \frac{25}{7} \times \frac{7}{25}$$

$$= 1$$

Exercise 8

The triangle AMC is obtained by increasing the height AM of triangle ABC by 15%. The triangle ANB is obtained by decreasing the base BC of triangle ABC by 8%. What is the ratio of the areas of triangles AMC and ANB?

Solution 8

Denote the area of triangle ABC by A. The area of triangle AMC is $1.15 \cdot A$. The area of triangle ANB is $0.92 \cdot A$. The ratio of the two areas is:

$$\frac{A_{\triangle AMC}}{A_{\triangle ANB}} = \frac{1.15 \cdot \cancel{A}}{0.92 \cdot \cancel{A}}$$
$$= \frac{115}{92}$$
$$= \frac{23 \times 5}{23 \times 4}$$
$$= \frac{5}{4}$$

The areas are in a ratio of $5 : 4$.

Exercise 9

The average of four numbers is x. The numbers are in a ratio of $2 : 3 : 5 : 8$. What is the range of the numbers as a function of x?

(A) $\dfrac{x}{2}$

(B) $\dfrac{3x}{4}$

(C) $\dfrac{4x}{3}$

(D) $\dfrac{2x}{3}$

Solution 9

Denote the numbers by $2a$, $3a$, $5a$, and $8a$. The range is $8a - 2a = 6a$.

Since their average is x:

$$\frac{2a + 3a + 5a + 8a}{4} = x$$

$$\frac{18a}{4} = x$$

$$\frac{9a}{2} = x$$

$$a = \frac{2x}{9}$$

The range is:

$$6a = 6 \times \frac{2x}{9} = \frac{4x}{3}$$

The correct answer is (C).

Exercise 10

After two successive price decreases, by 10% and 20% respectively, the price of a car is now $36,000$ dollars. What was the price before the rebates?

Solution 10

Denote the initial price of the car by C. The final price after the decreases is:

$$C \times 0.9 \times 0.8 = C \times 0.72$$

We can now solve for C:

$$
\begin{aligned}
C \times 0.72 &= 36000 \\
C &= \frac{36000}{0.72} \\
C &= \frac{3600000}{72} \\
C &= \frac{\cancel{9} \times 4 \times 100000}{\cancel{9} \times 8} \\
C &= \frac{\cancel{4} \times 100000}{2 \times \cancel{4}} \\
C &= 50000 \text{ dollars}
\end{aligned}
$$

Exercise 11

A number a is 20% of another number b. A number c is 20% of a and a number d is 60% of c. Which of the following is equal to the sum $a + b + c + d$? Check all that apply.

(A) $6.32 \cdot a$

(B) $1.264 \cdot b$

(C) $31.6 \cdot c$

(D) $52.\overline{6} \cdot d$

Solution 11

Express all numbers as a function of a:

$$
\begin{aligned}
a &= 0.2 \times b \\
c &= 0.2 \times a = 0.2 \times 0.2 \times b = 0.04 \times b \\
d &= 0.6 \times c = 0.6 \times 0.04 \times b = 0.024 \times b
\end{aligned}
$$

The sum of the four numbers is, as a function of b:

$$
0.2 \cdot b + b + 0.04 \cdot b + 0.024 \cdot b = 1.264 \cdot b
$$

110

The same sum is, as a function of a:

$$b = \frac{1}{0.2} \times a = 5 \times a$$

$$c = 0.2 \times a$$

$$d = 0.6 \times c = 0.6 \times 0.2 \times 1 = 0.12 \times a$$

$$a + b + c + d = a + 5 \cdot a + 0.2 \cdot 1 + 0.12 \cdot a = 6.32 \cdot a$$

As a function of c:

$$a = 5 \times c$$

$$b = 5 \times a = 25 \times c$$

$$d = 0.6c$$

$$a + b + c + d = 5 \cdot c + 25 \cdot c + c + 0.6 \cdot c = 31.6 \cdot c$$

As a function of d:

$$c = \frac{10}{6} \times d = \frac{5}{3} \times d$$

$$a = 5 \times c = 5 \times \frac{5}{3} \times d$$

$$b = 25 \times c = 25 \times \frac{5}{3} \times d$$

$$
\begin{aligned}
a + b + c + d &= \frac{5}{3} \cdot d \, (5 + 25 + 1) + d \\
&= \frac{5}{3} \cdot (30 + 1) \cdot d + d \\
&= 50 \cdot d + d + \frac{5}{3} \cdot d = 52.\overline{6} \cdot d
\end{aligned}
$$

All the choices are correct.

Exercise 12

15% of the number of girls in Amira's class is equal to 18% of the number of boys. What is the smallest number of students the class may have?

Solution 12

Denote the number of girls by G and the number of boys by B. We know that:

$$\frac{15}{100} \times G = \frac{18}{100} \times B$$

$$15 \cdot G = 18 \cdot B$$

$$5 \cdot G = 6 \cdot B$$

Since the numbers G and B must be integers, the lowest value of G must be 6 and the lowest value of B must be 5.

The smallest possible number of students in Amira's class is 11.

Exercise 13

The number of girls in Dina's class is equal to 80% of the number of boys. What percent of the number of girls is the number of boys?

Solution 13

Denote the number of girls by G and the number of boys by B. We know that:

$$G = \frac{80}{100}B$$

Then, we also have:

$$B = \frac{100}{80} \times G$$

$$B = \frac{5}{4} \times G$$

$$B = \frac{125}{100} \times G$$

The number of boys is 125% of the number of girls.

Exercise 14

Robot BeeToo tiles a floor using square tiles. Robot DeeFoo tiles an identical floor with tiles that have a side equal to 25% of the side of the tiles used by BeeToo. DeeFoo places tiles at a frequency that is 700% higher than that of BeeToo. Which of the following is true?

(A) In the same time, BeeToo can tile 2 times as much surface as DeeFoo.

(B) In the same time, DeeFoo can tile 2 times as much surface as BeeToo.

(C) It takes DeeFoo 4 times longer than BeeToo to tile the entire floor.

(D) It takes BeeToo 4 times longer than DeeFoo to tile the entire floor.

Solution 14

If the side of the tiles placed by DeeFoo is one quarter of the side of the tiles placed by BeeToo, then DeeFoo must place 16 times as many tiles to cover the same area. We know that DeeFoo places 8 tiles in the time BeeToo places 1 tile. It takes DeeFoo twice as long to tile an identical floor. The correct answer is (A).

Exercise 15

Alfonso, the grocer, sold 900 lbs of potatoes in June and 744 lbs of potatoes in July. What was the percent decrease in average daily sales between the two months?

Solution 15

The average daily sales in June were:

$$\frac{900}{30} = 30 \text{ lbs/day}$$

The average daily sales in July were:

$$\frac{744}{31} = 24 \text{ lbs/day}$$

113

The percent decrease was 20%:

$$\frac{24 - 30}{30} = -\frac{6}{30}$$

$$= -\frac{1}{5} = -20\%$$

Exercise 16

A delivery truck left the avocado farm to bring a shipment of avocadoes to Alfonso's grocery. The truck was supposed to reach the grocery at 9:00 AM, traveling at an average speed of 50 mph. However, the truck arrived at 10:30 AM because it only managed an average speed of 40 mph. What was the distance between the avocado farm and Alfonso's store?

Solution 16

The distance is the same in both scenarios. Denote the time it takes the truck to travel the distance at the expected speed of 50 mph by T. Then, the time it actually took the truck to travel was $T + 1.5$, in hours. The distance is the same in both cases:

$$50 \times T = 40 \times (T + 1.5)$$
$$(50 - 40) \times T = 40 \times 1.5$$
$$10 \times T = 40 \times 1.5$$
$$T = 4 \times 1.5$$
$$T = 6 \text{ hours}$$

The distance was $50 \times 6 = 200$ miles.

Exercise 17

Lila and Dina rode a tandem bike. For the first half of the time, they rode at an average speed of 3 mph. For the second half of the time, they rode at an average speed of 4.5 mph. What was their overall average speed?

Solution 17

Denote the total time they rode the bicycle by $2T$. The distance they covered was:

$$3 \times T + 4.5 \times T = 7.5 \times T$$

Their average speed was:

$$
\begin{aligned}
\text{Rate} &= \frac{\text{Distance}}{\text{Time}} \\[2mm]
&= \frac{7.5 \times T}{2 \times T} \\[2mm]
&= 3.75 \text{ mph}
\end{aligned}
$$

Exercise 18

Lila and Dina rode a tandem bike. For the first half of the way, they rode at an average speed of 3 mph. For the second half of the way, they rode at an average speed of 4.5 mph. What was their overall average speed?

Solution 18

Denote the total distance by $2D$. Half of the distance is D. The time it took them to cover the first half was:

$$T_1 = \frac{D}{3}$$

The time it took them to cover the second half was:

$$T_2 = \frac{D}{4.5}$$

The total time is:

$$T = \frac{D}{3} + \frac{D}{4.5} = \frac{7.5 \cdot D}{13.5} = \frac{15 \cdot D}{27} = \frac{5 \cdot D}{9} \text{ hours}$$

The average speed was:

$$
\begin{aligned}
\text{Rate} \; &= \; \frac{\text{Distance}}{\text{Time}} \\[2mm]
&= \; \frac{2 \cdot D}{\frac{5 \cdot D}{9}} \\[2mm]
&= \; \frac{2 \cdot D}{1} \times \frac{9}{5 \cdot D} \\[2mm]
&= \; \frac{18}{5} \\[2mm]
&= \; 3.6 \text{ mph}
\end{aligned}
$$

Exercise 19

A 4×6 rectangle, a 5×2 rectangle, and a 2×3 rectangle are placed together, without overlap, so that they form a figure with the smallest possible perimeter. After forming the figure, which of the following is closest to the percent decrease of the total perimeter?

(A) 18%

(B) 24%

(C) 28%

(D) 36%

Solution 19

The rectangles have a combined perimeter of:

$$2 \times (4 + 6) + 2 \times (5 + 2) + 2 \times (2 + 3) = 44 \text{ units}$$

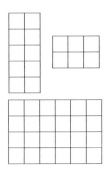

The smallest perimeter of the figure is 28 units. There are several configurations that result in this perimeter. One of them is:

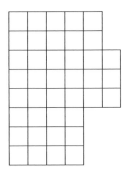

The percent change of the total perimeter is:

$$\frac{28 - 44}{44} \times 100\% \approx -36\%$$

and represents a decrease of approximately 36%.

CHAPTER 15. SOLUTIONS TO MISCELLANEOUS PRACTICE

Exercise 20

A truck and a car leave point P at 8:00 AM and travel towards point Q at 50 mph and 60 mph respectively. A motorcycle leaves point Q at 8:30 AM and travels towards point P at 60 mph on the same road. The motorcycle meets the car and, 12 minutes later, meets the truck. What is the distance between P and Q?

Solution 20

Denote the distance between the two points by D.

Denote the time elapsed from 8:00 AM by t. The time elapsed from 8:30 AM is $t - 0.5$, in hours.

Calculate the meeting time of the car and motorcycle:

$$
\begin{aligned}
60 \cdot t_1 &= D - 60 \times (t_1 - 0.5) \\
60 \cdot t_1 &= D - 60 \cdot t_1 + 30 \\
120 \cdot t_1 &= D + 30 \\
t_1 &= \frac{D + 30}{120}
\end{aligned}
$$

Calculate the meeting time of the truck and motorcycle:

$$
\begin{aligned}
50 \cdot t_2 &= D - 60 \times (t_2 - 0.5) \\
50 \cdot t_2 &= D - 60 \cdot t_2 + 30 \\
110 \cdot t_2 &= D + 30 \\
t_2 &= \frac{D + 30}{110}
\end{aligned}
$$

The difference between the meeting times is 12 minutes, or one fifth of an hour:

$$
\begin{aligned}
t_2 - t_1 &= \frac{D+30}{110} - \frac{D+30}{120} \\[2ex]
&= \frac{12D + 360 - 11D - 330}{10 \times 11 \times 12} \\[2ex]
&= \frac{D+30}{10 \times 11 \times 12} \\[2ex]
\frac{1}{5} &= \frac{D+30}{10 \times 11 \times 12} \\[2ex]
5 \times (D+30) &= 10 \times 11 \times 12 \\[1ex]
D + 30 &= 2 \times 11 \times 12 \\[1ex]
D + 30 &= 264 \\[1ex]
D &= 234 \text{ miles}
\end{aligned}
$$

Exercise 21

The sum of three numbers is 111. If we increase the first number by 150%, decrease the second one by 25%, and decrease the third one by 5, the results are equal. Find the numbers.

Solution 21

Denote the numbers by a, b, and c. If we increase a by 150%, we obtain $2.5 \cdot a$. If we decrease b by 20%, we obtain $0.8 \cdot b$. If we decrease c by 5, we obtain $c - 5$.

We know that:
$$2.5 \cdot a = 0.8 \cdot b = c - 5$$

and that:
$$a + b + c = 111$$

Since $c = 0.8 \cdot b + 5$ and $a = \dfrac{8}{25} \times b$:

$$\frac{8}{25} \times b + b + 0.8 \times b + 5 = 111$$

$$b \times \left(\frac{8}{25} + \frac{4}{5} + 1 \right) = 106$$

$$b \times \left(\frac{8 + 20 + 25}{25} \right) = 106$$

$$b = 106 \times \frac{25}{53}$$

$$b = 50$$

Then, $c = 0.8 \times 50 + 5 = 45$ and $a = \dfrac{8}{25} \times 50 = 16$.

Exercise 22

The base of a triangle increased by 15% and the corresponding height decreased by 10%. Which of the following is closest to the percent increase or decrease of its area?

(A) 3.5% increase

(B) 5% increase

(C) the area remains the same

(D) 0.5% decrease

Solution 22

The new height is:

$$H_{\text{new}} = 0.9 \cdot H_{\text{old}}$$

The new base is:

$$B_{\text{new}} = 1.15 \cdot B_{\text{old}}$$

The new area is related to the old area in the following way:

$$
\begin{aligned}
A_{\text{new}} &= \frac{B_{\text{new}} \times H_{\text{new}}}{2} \\[2mm]
&= \frac{B_{\text{old}} \times H_{\text{old}}}{2} \times 1.15 \times 0.9 \\[2mm]
&= 1.035 \times \frac{B_{\text{old}} \times H_{\text{old}}}{2} \\[2mm]
&= 1.035 \times A_{\text{old}}
\end{aligned}
$$

The area increased by 3.5%. The correct answer choice is (A).

Exercise 23

Lila pours 20 mL of milk in a bottle with 200 mL of coffee. Dina pours 20 mL of coffee in a bottle with 200 mL of milk. By how much percent is Lila's mixture more concentrated in coffee than Dina's mixture?

Solution 23

The concentration of coffee in Lila's mixture is:

$$C_L = \frac{200}{220} = \frac{10}{11}$$

The concentration of coffee in Dina's mixture is:

$$C_D = \frac{20}{220} = \frac{1}{11}$$

$$C_L = 10 \cdot C_D = C_D + 9 \times C_D$$

Lila's mixture is 900% more concentrated in coffee than Dina's mixture.

Exercise 24

Ali and Baba were counting their gold coins. Ali said to Baba: "Right now, you have twice as many gold coins as I have. If you give me 205 gold coins and if I give you 305 gold coins, then you will have four times as many gold coins as I have." How many gold coins did Ali have?

Solution 24

Denote the amount of money Ali had by M. Then Baba had $2M$. After the exchange, Ali would have $M + 205 - 305$ and Baba would have $2M - 205 + 305$ and the following would be true:

$$4 \times (M + 205 - 305) = 2M - 205 + 305$$

Solve for M to find the answer:

$$
\begin{aligned}
4 \times (M - 100) &= 2M + 100 \\
4M - 400 &= 2M + 100 \\
2M &= 500 \\
M &= 250 \text{ gold coins}
\end{aligned}
$$

Competitive Mathematics Series for Gifted Students

Practice Counting (ages 7 to 9)

Practice Logic and Observation (ages 7 to 9)

Practice Arithmetic (ages 7 to 9)

Practice Operations (ages 7 to 9)

Practice Word Problems (ages 9 to 11)

Practice Combinatorics (ages 9 to 11)

Practice Arithmetic(ages 9 to 11)

Practice Operations (ages 9 to 11)

Practice Word Problems (ages 11 to 13)

Practice Combinatorics (ages 11 to 13)

Practice Arithmetic and Number Theory (ages 11 to 13)

Practice Algebra and Operations (ages 11 to 13)

Practice Geometry (ages 11 to 13)

Practice Word Problems (ages 12 to 15)

Practice Algebra and Operations (ages 12 to 15)

Practice Geometry (ages 12 to 15)

Practice Number Theory (ages 12 to 15)

Practice Combinatorics and Probability (ages 12 to 15)

This is a series of practice books. With the exception of a few reminders, there are no theoretical explanations. For lessons, please see the resources indicated below:

Find a set of free lessons in competitive mathematics at www.mathinee.com. Addressing grades 5 through 11, the *Math Essentials* on www.mathinee.com present important concepts in a clear and concise manner and provide tips on their application. The site also hosts over 400 original problems with full solutions for various levels. Selectors enable the user to sort essentials and problems by test or contest targeted as well as by topic and by the earliest grade level they can be used for.

Online problem solving seminars are available at www.goodsofthemind.com. If you found this booklet useful, you will love the live problem solving seminars.

Made in the USA
San Bernardino, CA
16 May 2016